John Hampden Porter

Wild beasts

A study of the characters and habits of the elephant, lion, leopard, panther, jaguar, tiger, puma, wolf, and grizzly bear

John Hampden Porter

Wild beasts

A study of the characters and habits of the elephant, lion, leopard, panther, jaguar, tiger, puma, wolf, and grizzly bear

ISBN/EAN: 9783337303730

Printed in Europe, USA, Canada, Australia, Japan

Cover: Foto ©berggeist007 / pixelio.de

More available books at **www.hansebooks.com**

A STUDY OF THE CHARACTERS AND HABITS OF THE
ELEPHANT, LION, LEOPARD, PANTHER, JAGUAR,
TIGER, PUMA, WOLF, AND GRIZZLY BEAR

BY

J. HAMPDEN PORTER

ILLUSTRATED

NEW YORK
CHARLES SCRIBNER'S SONS
1894

TO

Captain John G. Bourke

U. S. ARMY

IN TOKEN OF FRIENDSHIP
AND IN REMEMBRANCE OF THE TIME
WHEN WE STUDIED TOGETHER

CONTENTS

	PAGE
THE ELEPHANT	1
THE LION	76
THE LEOPARD AND PANTHER	136
THE JAGUAR	175
THE TIGER	196
THE PUMA	257
THE WOLF	306
THE GRIZZLY BEAR	352

LIST OF ILLUSTRATIONS

The Elephant		*Frontispiece*
The Lion		*Facing page* 76
The Leopard		" " 136
The Jaguar		" " 175
The Tiger		" " 196
The Puma		" " 257
The Wolf		" " 306
The Grizzly Bear		" " 352

WILD BEASTS

THE ELEPHANT

THE elephant — "My Lord the Elephant," as he is called in India — takes precedence of other quadrupeds upon several counts. Among these appear conspicuously the facts that he belongs to an ancient and isolated family, which has no near relations occupying lower stations in life; likewise, that from time immemorial these creatures have been strong enough to do as they pleased. This latter circumstance more particularly ensured the sincere respect of mankind, and throughout the records of the race we find its members in distinguished positions. Ganesha, the Hindu god of wisdom, had an elephant's head, and *Elephas Indicus* was worshipped from Eastern China to the highlands of Central India. In Africa this species only escaped adoration because the natives of that country were incapable of conceiving any of those abstract ideas which the animal embodied. Wherever an elephant has existed, however, men have looked up to him, and as he was not carnivorous, it comported with human reason ing to extol the benevolence of a being who, if otherwise constituted, might have done so much harm.

Oriental, classic, mediæval, and modern superstitions cluster about the elephant. Pliny and Ælian often seem to be mocking at popular credulity. "*Valet sensu et reliquâ sagacitate ingenii excellit elephas,*" says Aristotle, and Strabo writes in the same strain. One might nearly as well take the verses of Martial for a text-book as seek information among those errors and extravagancies of antiquity which Vartomannus brought to a climax.

It is no longer said that elephants who, to use Colonel Barras' words ("India and Tiger Hunting"), "are practically sterile in captivity," are so because of their modesty, or that this is attributable to a nobleness of soul which prevents them from propagating a race of slaves. Men would now be ashamed to say they are monotheists, and retire to solitudes to pray. But so little of comparative psychology is known, and the side lights which other sciences throw upon zoölogy are so much disregarded, that no hesitation is felt at comparing them with human beings, or measuring the faculties and feelings of a beast by standards set up in civilized society.

The elephant is a social animal; in all herds the units are family groups where several generations are often represented, and when the larger aggregate dissolves, it separates into family groups again. With this statement, anything like unanimity of opinion among authorities upon elephants is at an end.

It is said that years bring moroseness upon elephants, and that any evil tendencies they exhibit in youth are aggravated by age. Apart from what may be exceptional in cases of this kind, the biological law is that the characteristic features of species, whether physical or mental, are not developed until maturity. Most of those who know these animals personally agree in the opinion that solitary males are commonly dangerous; and although the existence of "rogue elephants," who always belong to this class, has been denied, confirmatory evidence is too strong to be rejected. When some member of a group becomes separated from its relations and is lost, when a young bull is driven off for precocity, or an old tusker retires to solitude because he has been worsted in combat with a rival, the change of state cannot fail to be distressing, and the individual to deteriorate. At certain seasons male elephants often voluntarily abandon the society of females, but not usually of each other. When they grow old, there is more or less tendency towards seclusion in all bulls. Retirement, however, when prompted by age, apathy, or loss of the incitements towards association, is not at all like exile while physical powers and feelings are in force.

Ferocity is much more frequently met with in elephants than most people suppose; and as it is with these animals in a wild state, so is it also among those in captivity. There is no reason why a captured savage should spontaneously evolve adornments to his moral character because he is under restraint. A vicious brute is only restrained by fear, and this coercive influence continues just so long as apprehension is not overbalanced by passion.

Charles John Andersson ("The Lion and the Elephant") infers from the ease with which this animal accommodates itself to those requirements involved in domestication that its "natural disposition is mild and gentle." G. P. Sanderson ("Thirteen Years among the Wild Beasts of India") holds that "obedience, gentleness, and patience . . . are the elephant's chief good qualities."

Corse, speaking from his long experience in the elephant stables at Teperah and other places, states that they constantly exhibit a rooted animosity to other animals, and towards the keepers and helpers attached to them; while Colonel Julius Barras says, "all the old tuskers I have seen in captivity have killed one or two persons in the course of their career."

Passing from domesticated individuals to protected herds, Dr. Holub ("Seven Years in South Africa") found that on the Cape Town reservations they had "lost all fear of man, and had become excessively dangerous." Elephants in the government forests of Ceylon, where they are not exposed to attack from sportsmen, are described by Colonel James Campbell ("Excursions in Ceylon") as vicious and aggressive. On the other hand, neither Forsyth, Hornaday, Dawson, nor any other writers who were acquainted with the condition of animals similarly situated in India, have noticed that a like change has taken place among them.

It has been mentioned already that the existence of "rogue elephants" is denied; but everything that has ever been said about the race has likewise been denied. Andersson remarks of the solitary elephant that "instances innumerable are on record of his attacking travellers and others who had not offended him in any way." A tusker "in seclusion," observes Major Leveson ("Sport in Many Lands"), is always "morose, vicious, and desperately cunning." Leveson, Andersson, Campbell, Baker, Cumming, and Selous had ample opportunities for convincing themselves of the reality of rogues.

Speaking of the species on both continents, we may consider them as but little entitled to much of their reputation for harmlessness. Sir Samuel Baker ("The Rifle and Hound in Ceylon") gives it as his opinion that they are "the most dangerous creatures with which a sportsman can contend;" and W. T. Hornaday ("Two Years in the Jungle") takes the same view.

An elephant never exhibits the blind and senseless ferocity of a black rhinoceros. He is often fully as fierce, and far more destructive, but this disposition does not display itself in the same way. Both of these animals will, however, attack by scent alone. It is not meant that in elephants this conduct is customary; all that is intended is to substantiate the occurrence of such an act.

This animal's character is more completely evinced in the expression "My Lord the Elephant" than it could be by any description, however true and striking. Sanderson explains that the title is not given in reverence so much as in fear. The native attendants upon elephants, he observes, have little respect for their intelligence, but a lasting apprehension of what may at any time happen to themselves.

It is generally said that while male elephants are free they never become "must," and, therefore, that this temporary delirium arising from interference with natural functions, cannot be the cause of those extreme cases of viciousness which occasionally make a tusker the scourge of a whole district. Whether "must" or not, these brutes are sometimes mad, and among other examples that might be given, Sir Samuel Baker's description of a "tank-rogue," — shot by himself in Ceylon, — portrays too faithfully the familiar symptoms of mania to leave any doubt about the animal's condition.

This fierce beast had committed many murders, — killing people without any provocation; lying in wait for them; stealing towards those places he knew to be frequented; and apparently devoting all his energies to the destruction of human life. From the first moment at which he was seen all his actions betokened insanity. Baker never suspected the true state of the case, but he watched this elephant for some time, and carefully noted his conduct, — his wild and disordered mien, his aimless restlessness, and causeless anger; all the features which form the characteristic physiognomy of mania.

Extremely dangerous elephants are not, however, always insane. There is no need to argue mental alienation in order to account for acts which vice of itself is fully competent to explain. The beast's strength is enormous, its bulk greatest among land animals, its offensive weapons and general capability of doing harm are unequalled. Of these facts the creature itself must be conscious, and it never exhibits the darker side of its character without showing that it is so.

This leads to a question that has been considerably disputed, and concerning which many opinions have been recorded — all dogmatic, and most of them contradictory. Suppose that a homicidal elephant catches a fugitive whom he pursues, how does he kill him, and is he invariably destroyed? The subject stated does not amount to much in itself, but some points will appear in the course of a brief inquiry into it that merit attention. All writers who held to the instinctive hypothesis, and imagined that brutes only acted in a predetermined way, have taken exclusive views of this matter. When a man is overtaken by an elephant many say he is always killed. Sanderson, for example, says so. Captain Wedderburn was killed. Professor Wahlberg was killed. Everybody is killed; it cannot be otherwise. Nevertheless, Colonel Walter Campbell ("The Old Forest Ranger") saw a companion emerge from beneath the feet of a rogue elephant, and Major Leveson and Major Blayney Walshe ("Sporting and Military Adventures in Nepaul") relate the incidents of like cases. Henry Courtney Selous ("A Hunter's Wanderings in Africa") lived to tell how this same good fortune attended himself; and Lieutenant Moodie was actually trampled in the presence of several witnesses, and yet, although considerably injured, escaped with his life.

These were, of course, very unusual instances, and it is undeniable that most people whom elephants catch are killed. But how? Pressed to death with one of the animal's forefeet, one authority declares; with both of them, another insists; kicked forwards and backwards between the hind and front legs till reduced to a pulp, maintains a third; transfixed with the tusks, kneeled upon, walked over, dismembered, others protest, as if any mode of putting a man to death, except that particular one which they had determined to be the natural, usual, and, so to speak, proper method, would be a singular departure from the course an elephant might have been expected to pursue.

Sir Emmerson Tennant ("Ceylon"), who has made as many mistakes about these animals as can anywhere be found gathered together in one place, is certain the tusks are never used offensively. He, in fact, shows that it is physically impossible that they should be. According to him these appendages are probably auxiliary to the animal's food supply, but for the most part useless. Nobody, however, ever saw a pair of these developed front teeth that were symmetrical; one is invariably more worn away than the other on account of its having been used by preference in digging up roots, bulbs, etc. With respect to their employment as weapons, Selous states that "when an elephant overtakes his persecutor [a man, that is to say], he emits scream after scream in quick succession, all the time stamping upon and ventilating his adversary with his tusks." That these are "most formidable weapons," remarks Sanderson, is recognized by the animals themselves. "Tuskers always maintain the greatest discipline in a herd. . . . Superiority seems to attach to one or the other in proportion to the size of the tusks;" and in the combats between bull elephants which he witnessed "one was often killed outright." Further, when a male has only one tusk, as not unfrequently happens, this is obviously more effective than both would be, and in that event, Sanderson adds, "he is the terror of an elephant corral . . . its undisputed lord." The weak point in Sir Emmerson Tennant's demonstration of the mechanical impossibility of using those parts, on account of the angle at which they are set in the jaw, is due to his having overlooked the fact that an elephant can move his head. Emin Pasha ("Collection of Journals, Letters, etc.") reports that he saw a soldier in Central Africa who had been desperately wounded by a thrust from an elephant's tusk. It was the accident of being struck by the side of one instead of its point that enabled Colonel Barras to get off with his life; and Sir Samuel Baker relates the death of Mr. Ingram, who was transfixed. These animals have no special way of inflicting death, though most commonly this is caused by trampling. All the modes enumerated are vouched for by witnesses whose evidence there is no reason to doubt, and this clash of opinion is only one of the many outgrowths of that strange superstition by which brutes are represented to act uniformly in consequence of their unvarying mental constitution. Nothing, for instance, even among the best authorities, is more frequently met with than the point-blank assertion that an elephant never strikes with its trunk. Yet Andersson ("Lake N'gami") was nearly killed in this way. General Shakespear saw his gun-bearer struck down, and Sir James E. Alexander ("Excursions in Africa") describes its use as a means of offence. There are many reasons why this organ should not be thus employed habitually, but there is no cause which would prevent it from being applied in this manner when the animal himself, who is much the best judge, thought proper to do so.

The effect upon these species of those general influences which are exerted by social life may be inferred from the existence of their coherent family groups, from the protracted period during which maternal guardianship is continued, and the baneful results that solitude brings about. Still there seems to be little doubt that Green, Moodie, and Pollok represent the best opinion in saying that sympathy is less active in elephants than it is in many animals whose moral qualities have usually been considered as greatly inferior to theirs. "I have never known an instance," remarks Sanderson, "of a tusker undertaking to cover the retreat of a herd."

Although elephants are often hysterical, and always nervous, discipline effects great changes in their ordinary conduct. At the same time, they can rarely be trusted. Sir Samuel Baker states ("Wild Beasts and Their Ways") that he had never ridden but "one thoroughly dependable elephant," and most tiger-hunters say the same.

Elephants are without ideals of any kind. They cannot be influenced by superstitions, and it is useless to explain their excellencies and defects by reference to a descent of which we know nothing, or to assume that transformations may be effected by means of an education that always begins *de novo*, and is in itself superficial and incomplete in the highest degree. Foreknowledge of those consequences entailed by misbehavior no doubt prompts most of the acts that are attributed to industry, magnanimity, friendliness, and forbearance, as attention to their keeper's directions explains the usual manifestations of intellect that have been so much admired.

Those who know them best think that elephants, as Sanderson expresses it, are "wanting in originality," so that when an unusual emergency occurs they feel at a loss. It is true that life is in some respects comparatively simple with these animals, and that its necessities neither involve the same constructions, nor require a like care with that imposed upon many others. But in those directions in which the struggle for existence engages their powers energetically they display considerable capacity, though not of the highest brute order. Colonel Pollok ("Sport in British Burmah") says, "if Providence has not given intellect to these creatures, it has given them an instinct next thing to it. . . . Providence has taught them to choose the most favorable ground, whether for camping or feeding, and to resort to jungles where their ponderous bodies so resemble the rocks and dark foliage that it is difficult for the sportsman to distinguish them from surrounding objects; whilst their feet are so made that not only can they tramp over any kind of ground, whether hard or soft, rough or smooth, but this without making a sound.

"Some of their camping-grounds are models of ingenuity, surrounded on three sides by a tortuous river, impassable by reason either of the depth of water, its precipitous banks, quicksands, or the entangling reeds in its bed; while the fourth side would be protected by a tangled thicket or a quagmire. In such a place elephants would be in perfect safety, as it would be impossible for them to be attacked without the attacking party making sufficient noise to put them on the alert.

"Their method of getting within such an enclosure is

also most ingenious. They will scramble down the bank where the water is deepest, and then, after either wading or swimming up or down stream, ascend the opposite bank a good half-mile or more from where they descended, thereby doubly increasing the difficulty of following them."

Many animals rival elephants in those respects described, and a few surpass them. All that they do has been too much exaggerated, and their unquestionable sagacity loses much of its point by being unduly exploited.

Relative complexity of structure in brain and mind is in no way more strongly marked than by the ability to suppress emotion. This is not the highest characteristic of an evolved organism, but it is one that no being which is not of a high grade can possess. When a captive elephant, often without any provocation, makes up its mind to commit murder, nothing can exceed the patience with which the animal awaits an opportunity, except its power of dissimulation. How it regards the contemplated act, what thoughts and feelings are agitated while brooding over its accomplishment, we do not know, but the history of many such cases has been fully given, and of the behavior displayed under these circumstances we can speak with certainty.

Generally elephants kill their attendants, as being those most likely to give offence. An antipathy is, however, sometimes conceived against some casual acquaintance, whose efforts to ingratiate himself have only inspired the creatures with a causeless hatred. It is the fashion to say that homicide by these beasts always indicates that they have been injured. People endow elephants with an exaggerated form of the sensitive pride belonging to human character, and, through some unexplainable process of thought, reconcile its coexistence with the malignant temper of a murderous brute. The way in which one of their attendants talks to an elephant whom he suspects is strange enough. This man despises his intellect, and knows his character thoroughly. " Have I ever been wanting in respect? *Astagh-fur-Ulla*. God forbid! Let my Lord remember how yesterday at bathing-time he was placed under a tree, while that son of Satan, Said Bahadur, stood in the sun. Who has provided your highness with sugar-cane, and placed lumps of goor between your back teeth? I represent that this, oh, protector of the poor, it was my good fortune to do. Hereafter I will deprive those unsainted ones about you of their provisions and bestow them upon you." That is the way a Hindu talks, hoping to mollify the animal.

Certain traits in animals have come to be accepted as peculiarly significant of their respective grades; parental affection, for example. The male elephant is as nearly as possible without a trace of this feeling, but his polygamous habits account to a great extent for the deficiency. It is a quality which greatly preponderates in females of most species, and in one so elevated we might expect to find that this, as Buffon asserts, was a prominent trait. Frederick Green informs us, however, that "the female elephant does not appear to have the affection for her offspring which one would be led to suppose," and his view is very far from being singular. The author has not found any justification in facts for Buffon's assertion to the contrary. Doctor Livingstone ("Travels and Researches in South Africa") reports the case of a calf elephant whom its mother abandoned when attacked, and Sir W. Cornwallis Harris ("Wild Sports in Southern Africa") says that a young animal of this kind if accidentally separated from its mother forgets her instantly, and seeks to attach itself to the nearest female it can find. Sanderson observes in this connection that "while the female evinces no particular affection for her progeny, still, all the attention a calf can get is from its own mother."

. G. Macloskie ("Riverside Natural History") states that "elephants are well disposed towards each other in aggregation." Evidently such must be the case, or they could not live together. Their gregarious habits imply an average friendliness.

While, however, their ordinary temper may, or rather must, be as stated, leadership in herds, when this is not held by a tuskless male or "some sagacious old female," whose abilities their companions are intelligent enough to understand, is settled by combat, and maintained in the same way. Moreover, bull elephants often quarrel and fight desperately in the free state, and it is said by one or two observers (Drummond particularly) that when herds intoxicate themselves, as they do upon every opportunity, with the *Um-ga-nu* fruit, they exhibit scenes of riot and violence which cannot be matched on earth. Captive tuskers in elephant stables are always at feud with some other animal, and all their inmates quarrel upon small provocation. Recently-captured elephants that have not been removed from the corral frequently attack each other, and when some lost or exiled wanderer attempts in his distress and loneliness to join another band, its champion at once assails him.

There is one detestable trait, not uncommon among many species, and shared by a portion of savage mankind, which elephants do not display. They never destroy injured or disabled animals of their own kind. On the contrary, when sympathy does not involve self-sacrifice, they sometimes (not always by any means) show that they are not without the feeling, and this conclusion seems to be quite capable of resisting all the destructive criticism that can be brought to bear upon it.

Wild beasts have usually been written about both carelessly and dogmatically. Men, for the most part, no doubt unconsciously, speak of them as if they knew what it is impossible that they should know; and it is difficult to banish the suggestion that many of our prevailing opinions are in fact survivals from savagery. Public feeling towards elephants is undoubtedly swayed by their size, and by involuntary apprehension. We are struck by the contrast between the animal's placid appearance and those powers it embodies. In short, people do not study elephants, or reason about them; they feel in a modified form those original impressions which operated upon their remote ancestors. Hence, in great measure probably, Buffon's ipse dixit, "*dans l'état sauvage, l'éléphant n'est ni sanguinaire, ni féroce, il est d'un natural doux, et jamais il ne fait abus de ses armes, ou de sa force.*" It is not so much the verbal statement that need be objected to in this sweeping assertion, as the spirit in which it is made. More is implied than said, and the implication is that an elephant is self-controlled by sentiments that are as foreign to its mind as a pair of wings would be to its body. A wild beast, which while free to follow its own devices and desires, does not conduct itself like a wild beast, is an impossibility in actual life.

Sanderson supposes that "all catching elephants"— the trained ones used in securing captives—"evince the greatest relish for the sport." This is a mild way of putting Sir Emmerson Tennant's opinion that they show a decided satisfaction, a malignant pleasure, such as Dr. Kemp ("Indications of Instinct") describes, in the misfortunes of their fellows. Now in what way Sanderson discovered that this state of mind existed cannot be divined, for he gives it as the result of his own direct observations, that "the term decoy is entirely misapplied to tame elephants catching wild ones, as they act by command of their riders, and use no arts. . . . The animal is credited with originating what it has been taught, with doing of itself what it has been instructed to do. . . . I have seen the cream of trained elephants at work . . . in Bengal and Mysore: I have managed them myself under all circumstances . . . and I can say that I never have seen one display any aptitude for dealing undirected with an unexpected emergency." Since he then believes them to be incapable of showing this "relish" by their actions, since he has never known them to do anything of themselves on these occasions, in what way did he find out how they felt?

All those who speak from experience concur in representing a hunted elephant who does not or cannot escape, as superlatively dangerous. This is not only attributable to the fact that he is then extremely fierce and determined, but also to his undoubted ability to use the great powers of attack and defence he possesses. The animal is capable of considerable speed for a short distance, but it is not possible for him to prolong effort to any great extent.

Selous asserts that no large creature, except a rhinoceros, matches the elephant in its activity upon rough ground. "They can wheel like lightning," says Baker; or, as Andersson expresses it, "Spin round on a pivot." Captain J. H. Baldwin ("Large and Small Game of Bengal") describes their performances upon hillsides as very remarkable.

Captain James Forsyth informs us of the ease and celerity with which they move over a broken surface. Inglis ("Work and Sport on the Nepaul Frontier") relates the dexterity and quickness of these ponderous beasts in crossing gullies that seem impassable. There is probably no animal safer to ride over a dangerous mountain road. Nervous as he is, his intelligence acts through a brain well enough organized to warn him against the consequences of carelessness. A horse will dash himself to death getting out of the way of a swaying shadow or whirling leaf, and on many journeys nobody thinks of mounting one; but the elephant's prudence, if not his courage, is, as a rule, to be relied upon.

It has somewhat arbitrarily been decided upon that an elephant can travel at the rate of fifteen miles an hour for a few hundred yards, and no faster. Its gait has been similarly settled by several authorities. Dr. Livingstone declares that the animal's "quickest pace is only a sharp walk." Sanderson modifies this statement by saying that the rapid walk "is capable of being increased to a fast shuffle." He adds the information that "an elephant cannot jump . . . can never have all four feet off the ground at once . . . and can neither trot, canter, nor gallop." Joseph Thomson, however ("Through Masai Land"), saw one of these animals which he had wounded on the plateau of Baringo, "go off in a sharp trot," and Colonel Barras, while beating a clump of bushes for a wounded tiger, rode his Shikar tusker Futteh Ali almost over the concealed brute; whereupon says Barras, "he spun round with the utmost velocity and fled at a rapid gallop. The pace was so well marked that it would be useless, as far as I am concerned, for any one to say that it was mechanically impossible for an elephant to use this gait. To such learned objectors I would point out the fact that impossibilities are of daily occurrence, and would further beg them to suspend judgment till they have sat on an elephant's neck with an enraged tiger roaring at his heels." Much the same restriction has been placed by some naturalists upon the camel's paces. Nevertheless, Sir Samuel Baker and G. C. Stout were convinced that they had seen camels trot, and the author is quite as certain as Colonel Barras could possibly be that he has known them to gallop.

It has been the fashion to praise these animals indiscriminately. Among other things the silence maintained by so bulky a creature, and the noiselessness of its movements, are mentioned as evidences of great sagacity. An elephant, however, cannot make a noise with its feet except by kicking something out of the way or breaking it; their formation renders its tread, under ordinary circumstances, inaudible. The body also being elliptical in its long diameter, passes through undergrowth, when the animal is moving slowly, like a vessel through water. Further, obstacles that do not offer too much resistance are put aside easily by the trunk, which has all those varieties of motion that about fifty thousand sets of muscles can confer. More than this, quietness is not necessarily a mark of caution, foresight, or self-restraint, and some of the wariest creatures in existence are by no means quiet. As a matter of fact, if not alarmed or asleep, — in which case he snores in a manner conformable with his size, — the elephant is one of the noisiest of wild beasts. A perpetual crashing accompanies both individuals and herds while feeding, and in hours of repose they frequently trumpet, their deep abdominal rumble is often heard, and sounds expressive of contentment or dissatisfaction constantly break the silence of the forest.

When danger is apprehended, if they do not dash away "with the rush of a storm," elephants are apt to remain motionless for a time, while straining their most perfect senses — those of hearing and smell — in order to ascertain its character and proximity, or one or more may advance cautiously in order to see. Having done this, they depart as secretly as possible, and in the way mentioned, but why anybody should wonder that these creatures, whose sagacity is considered to be so extraordinary, do not move off abreast instead of in single file, as is their custom, and thus voluntarily encounter the greatest amount of resistance, and ensure the most disturbance, it is not easy to understand. In all measures relating to evasion, as contradistinguished from precaution, these beings occupy an inferior position: their color makes them nearly indistinguishable in those places they mostly occupy, and the footfall is naturally noiseless, but they employ none of those arts in which many species are expert, and do not even confuse their trail. This deficiency in cunning cannot be accounted for by the offhand explanation that the elephant, conscious of his strength, has no need to conceal himself. He has fully as much, if not more reason to do so, than many other animals, and the experience by which the latter have profited has been common to them all.

Those inferences which have oftentimes been drawn from the social life of elephants will scarcely stand the tests furnished by sociology. "A herd of elephants," observes Leveson, "is not a group that accident or attachment may have induced to associate together, but a family," between whose members "special resemblances attest their common origin." Reasoning from statements like this, it is concluded that results accrue from an aggregation of relatives similiar to those which obtain in human families;—that they are, in effect, groups of the same kind, saved from disruption and made amenable to improvement by mutual aids, forbearances, affections, and distributions of office. But those resemblances discoverable do not warrant the comparison.

What we know of social groups among elephants is that they are unlike those formed by mankind. It is doubtful whether the family, properly so-called, primarily exists in human society, and whether it is not a later combination instituted upon the basis of common possessions. Starcke ("The Primitive Family") holds that such is the case, and his view has not been shown to be incorrect. If this is true, to compare these congregations is to place lower animals by the side of human beings who have already taken an important step in advance. As a matter of fact, the qualities by which such groups are united among mankind, are to a great extent wanting with elephants. They cannot be wholly absent, but they are inconspicuous and obscured by disaggregative tendencies. As life advances, age does not bring with it a fruition of those tendencies upon which family ties depend; time only tends to exaggerate everything that is unsocial in the brute's nature.

Many conclusions respecting the intellect and emotional character of elephants have been drawn from untrustworthy anecdotes. It is in an uncritical spirit that Professor Robinson ("Under the Sun") reports the behavior of that famous tusker who bore the imperial standard on some old Mogul-Mahratta battle-field. The day had gone against his side, the color-guard was scattered, broken squadrons swept past the elephant, and his mahout was dead. He stood fast, however, and finally the retreating forces rallied around him, and the field was retrieved. Taken literally, his conduct amounted to this; namely, that his keeper whom he was accustomed to obey, ordered him to stand still, and he did so. Of course this animal possessed unusual nerve, but what else did he have? The high sense of duty Professor Robinson has discovered; heroic self-sacrifice that kept him, like the unrelieved Roman sentinels at Pompeii, on his post to the last? There is just the same reason for thinking so as there is for giving to the riderless horses who galloped with the Light Brigade towards the Russian guns at Balaklava, the sentiments of those soldiers who made that gallant but useless charge.

So it is with all instances of a like character. There are many more accounts of the elephant's cowardice than of its courage, and it is notoriously untrustworthy in war. Some are braver than others, but as soon as we attempt to find out from the literature of this subject which are the bravest, — young or old, male or female, trained or untrained, wild or tame, — hopelessly contradictory statements crowd upon us from all sides. The highest, the most complete, the severest discipline this beast receives is in the hunting-field, and Colonel MacMaster expresses the general tenor of opinion upon its results in saying, "I have never known an elephant who could be depended upon for dangerous shooting." As a class these animals are liable to panic, easily confused, and often become imbecile on account of nervous agitation. It is not uncommon to see a tusker fly screaming with fear from the skin of a tiger which he has seen taken off, or to have him bolt from its dead body when that is instantly recognized as harmless by the jungle crow, pea-fowl, or monkey. Being extremely afraid of bears for some unknown reason, and nearly idiotic when frightened, an elephant may attack the hunter who has just stepped off his back into a tree, thinking that he has been suddenly transformed into a brute of this kind. But from all appearances some of them like to hunt, and when well broken and in good health, their prompt and intelligent obedience, their display of natural powers of several kinds, and the firmness with which they confront danger and bear pain, are wonderful.

Neither the man on his back nor the elephant himself is by any means secure against fatal results when a tiger charges home. Shikar animals, nevertheless, often do everything that is required of them admirably. The difficulty is that the best elephants cannot be counted upon. A tusker, whose scars speak for themselves, is as likely as not, says Colonel MacMaster, "to bolt from a hare or small deer, or quake with fear when a partridge or peafowl rises under his trunk."

The following narrative by Captain James Forsyth ("The Highlands of Central India") illustrates some of the foregoing criticisms very well: —

"It was in 1853 that the two brothers N. and Colonel G. beat the covers" of Bétúl, near the village of Bhádúgaon, "for a family of tigers said to be in it. One of the brothers was posted in a tree, while G. and the other N. beat through on an elephant. The man in a tree first shot two of the tigers, and then Colonel G. saw a very large one lying in the shade of a bush and fired at it, on which it charged and mounted the elephant's head. It was a small female elephant, and was terribly punished about the trunk and eyes in this encounter, though the mahout (a bold fellow named Rámzán, who was afterwards in my own service) battered the tiger's head with his iron driving-hook so as to leave deep marks in the bones of his skull. At length he was shaken off, and retreated; but when the sportsmen urged in the elephant again, and the tiger charged as before, she turned round, and the tiger catching her by the hind leg fairly pulled her over on her side. My informant, who was in the howdah, said that for a time his arm was pinned between it and the tiger's body, who was making efforts to pull the shikari out of the back seat. They were all, of course, spilt on the ground with their guns, and Colonel G., getting hold of one, made the tiger retreat with a shot in the chest. The elephant had fled from the scene of action, and the two sportsmen then went in at the beast on foot. It charged again, and when close-to them was finally dropped by a lucky shot in the head. But the sport did not end here, for they found two more tigers in the same cover immediately afterwards, and killed one of them, making four that day. The worrying she had received, however, was the death of the elephant, which was buried at Bhádúgaon,—one of the few instances on record of an elephant being actually killed by a tiger."

There is no way in which the intellect, moral attributes, temper, receptive power, and adaptability of elephants can be decided upon *en masse*. An animal of this kind will tend his keeper's infant with a solicitude which seems to justify all that has been said of his benevolence; he will also watch for an opportunity to kill its father with a patience and self-command that are more significant still.

In the latter event the motive (hatred) displays itself, and the manner in which the design is carried out can be studied ; but with respect to the determining causes of conduct in the first instance we know nothing. An intelligent animal has been told to do something which it understands, and does it to the best of its ability. That is all the facts warrant us in saying.

One way of estimating the degree of feeling in any case is to measure the actions that express it by what they cost the individual who performs them. An elephant's opportunities for displaying self-abnegation can be but few, and most of those voluntary deeds upon which his reputation rests require little or no self-forgetfulness. In the hunting-field he is under coercion. A hunted elephant, however, is not in this position, and it is in its conduct that we notice such examples of this kind of behavior as may be regarded in the light of cases in point. Elephants — females most frequently — sometimes fight in defence of their associates when they themselves are not directly attacked. Both sexes have been occasionally known to give assistance to each other when they might have been killed in doing so. But for the most part they are very far from acting in this way. Fishes, reptiles, birds, together with a large number of land animals, have fully equalled elephants in everything they have done in this direction. Much has been said of the affection an elephant feels for the person who feeds and tends it, of the care, consideration, respect, and obedience it renders to a being whose superiority this amazing brute recognizes. Nevertheless, it is most probable that this individual had better be anywhere else than within reach of its trunk if there is a probability of the animal's getting bogged, for the chances are that he will be buried beneath its feet for a support.

This is not said with the intention of disparaging those good qualities which elephants possess. It must be plain from what has gone before that nothing else was to be expected. Except in the way of patient dissimulation, it would be difficult to show that when these animals take to evil courses they display more ability in perpetrating crime than many others. The consequences of vice in them are apt to be serious, and thus attract attention; but so far as cunning, foresight, and invention are called into play, they do not distinguish themselves, and those tragedies with which their names are associated seem to be more particularly marked by violence, ferocity, and rapidity of execution. Furthermore, it is well known that cerebral structure in these species is not of a high type; and with regard to its organization we know nothing.

If we now follow this largest of game into its native haunts, and note those experiences by which its pursuit is attended, what has been said with reference to the habits and character of elephants will, in the main, be found to rest upon good evidence. The outlook will be quite different according to where the animals are found. In India elephants live almost altogether in forests, while in Africa this is not the case. A hunter on the "Dark Continent" may also ride; quite an advantage in escaping a charge, and also in following a beast who, when frightened, frequently goes forty miles at a stretch. Dogs can always divert this creature's attention from the man who is about to kill him. The barking of a few curs about his feet never fails to make an enraged elephant forget the object of attack.

Sir Samuel Baker ("Wild Beasts and their Ways") and Colonel Pollok ("Sport in British Burmah") have described at length the most vulnerable points in the body and head, but sporting stories and details, except in so far as they illustrate temper and traits of character, are beside the purpose here. It may be said, however, that the forehead shot, so constantly made in India, cannot be resorted to with an African elephant. It has been tried a great many times, and there are only two or three instances on record where the animal has been killed. This is due to a difference of conformation in the skull, in the position of the brain, and to the manner in which this elephant holds its head in charging, says F. C. Selous ("Travel and Adventure in South East Africa").

Without going into anatomical details, it may be said that an African is about a foot taller than an Indian elephant, his ears are much larger, his back is concave instead of convex, and the tusks are much heavier and longer. Their position in the jaw also differs; they converge in passing backwards and upwards into the massive processes in which they are set, so that their roots, and the masses of bone and cartilage which form their sockets, effectually protect the brain, which lies low behind the receding forehead.

Speaking of hunting on horseback, W. Knighton ("Forest Life in Ceylon") mentioned it as a well-known fact that "the elephant has an antipathy towards a horse." "A solitary traveller is perfectly safe while mounted" he remarks. To the best of the author's knowledge and belief, the fact is directly the other way. Horses, until accustomed to their sight and odor, fear elephants, but the latter care nothing about them. They have never been known to hesitate in attacking hunters in the saddle. The Hamran and Baggara Arabs on the Upper Nile and its tributaries nearly always meet them in this manner. The only weapon used by these aggageers, or sword-hunters, is a long, heavy, sharp, double-edged Solingen blade. Three men generally hunt together, and their method of procedure shows how well they know the elephant's character.

Having found the fresh spoor of an old bull whose tusks are presumably worth winning, they track it to its resting or feeding place, and approach with no other precaution than is necessary to keep their quarry from taking refuge in some mimosa thicket where their swords cannot be used. When possible, the animal, who appreciates the situation perfectly, and knows all about sword-hunters, always makes itself safe in that way. If no cover is within reach, the elephant backs up against a rock, a clump of bushes, bank, or anything that will guard it in the rear, and awaits its enemies with that peculiarly devilish expression of countenance an elephant wears when murderously inclined. Supposing the aggageers to be three in number, and mounted, — two of them close slowly in upon his flanks, while the third — the lightest weight, on the most active and best broken horse — gradually approaches in front. There stands the elephant with cocked ears and gleaming eyes, and the Arab slowly drawing nearer, sits in his saddle and reviles him. Finally, what the Hamrans or Baggaras knew from the first would happen actually takes place. The elephant forgets everything, and dashes forward to annihilate this little wretch who has been cursing and pitching pieces of dirt at him. Then the horse is whirled round, and keeping just out of reach of his trunk, its rider lures the enraged animal on. As soon as he starts, those riders on his quarters swoop down at full speed, and when the one on his left comes alongside, he springs to the ground, bounds forward, his sword flashes in the air, and all is over. The foot turns up in front, in consequence of cutting the tendon that keeps it in place, and its blood rapidly drains away through the divided vessels until the animal dies.

That "the reasoning elephant," of whom Vartomannus ("*Apud Gesnerum*") exclaims in terms that have been repeated for nearly two thousand years, "*Vidi elephantos quosdam qui prudentiores mihi vidabantur quàm quibusdam in locis hominis,*" should have thus relinquished his advantages, abandoned an unassailable position, and knowing the consequences, rushed upon destruction in this way, is deplorable, and the worst of it is that he always does this. The intellect of which Strabo calmly asserts that it "*ad rationale animal proxime accedit,*" is never sufficient to save him. Probably, however, this conduct might appear to be more consistent, if instead of trusting to these very classical but perfectly worthless opinions, we looked upon it from the standpoint which Sanderson's description affords. "Though possessed of a proboscis which is capable of guarding it against such dangers, the elephant readily falls into pits dug to receive it, and which are only covered with a few sticks and leaves. Its fellows make no effort (in general) to assist the fallen one, as they might easily do by kicking in the earth around the edge, but fly in terror. It commonly happens that a young elephant tumbles into a pit, near which its mother will remain till the hunters come, without doing anything to help it; not even feeding it by throwing in a few branches. . . . Whole herds of elephants are led into enclosures which they could break through as easily as if they were made of corn stalks . . . and which no other wild animal would enter; and single ones are caught by their hind legs being tied together by men under cover of tame elephants. Animals that happen to escape are captured again without trouble; even experience does not bring them wisdom. I do not think that I traduce the elephant, when I say that it is, in many things, a stupid animal."

Baldwin, Harris, and a few other authorities, report that elephants are sometimes attacked by the black rhinoceros, but otherwise they have no foes except man. In Sir James Alexander's account ("Excursion into Africa") of the manner in which these beasts attempt to defend themselves against the charge of an enemy of this kind, it is implied that the trunk is habitually used offensively. "In fighting the elephant," he observes, the two-horned black rhinoceros, for no white rhinoceros ever does this, "avoids the blow with its trunk and the thrust with its tusks, dashes at the elephant's belly, and rips it up." Quite a number of writers have derided and denied statements of this nature, and if it were not that they have likewise scouted everything which they did not see themselves, their dissent might have more weight than it has. Everybody knows that the species of rhinoceros spoken of are of all wild beasts the most irritable, aggressive, and blindly ferocious; that they will, as Selous asserts, "charge anybody or anything." Apart from the question whether this kind of combat ever takes place, or what the result would be if it did, so many reasons exist why the trunk should not be used like a flail, as here represented, that good observers have failed to recognize the fact that it sometimes is so employed. At all events, in face of various assertions to the effect that it never strikes with its trunk, we find Andersson nearly killed in this manner. He was shooting from a "skärm"; that is to say, a trench about four feet deep, twelve or fifteen long, and strongly roofed except at the ends. This hiding-place and fortification occupied "a narrow neck of land dividing two small pools"—the water-holes of Kabis in Africa. "It was a magnificent moonlight night," and the hunter soon heard the beasts coming along a rocky ravine near by. Directly, "an immense elephant followed by the towering forms of eighteen other bulls" moved down from high ground towards his hiding place, "with free, sweeping, unsuspecting, and stately step." In the luminous mist their colossal figures assumed gigantic proportions, "but the leader's position did not afford an opportunity for the shoulder shot," and Andersson waited until his "enormous bulk" actually towered above his head, without firing. "The consequence was," he says "that in the act of raising the muzzle of my rifle over the skärm, my body caught his eye, and before I could place the piece to my shoulder, he swung himself round, and with trunk elevated, and ears spread, desperately charged me. It was now too late to think of flight, much less of slaying the savage beast. My own life was in the most imminent jeopardy; and seeing that if I remained partially erect he would inevitably seize me with his proboscis, I threw myself upon my back with some violence; in which position, and without shouldering the rifle, I fired upwards at random towards his chest, uttering at the same time the most piercing shouts and cries. The change of position in all probability saved my life; for at the same instant, the enraged animal's trunk descended precisely upon the spot where I had been previously crouched, sweeping away the stones (many of them of large size) that formed the front of my skärm, as if they had been pebbles. In another moment his broad forefoot passed directly over my face." Confused, as Andersson supposed, by his cries, and by the wound he had received, the elephant "swerved to the left, and went off with considerable rapidity."

Of course, taking this narrative literally, it may be said that it is not an illustration of the point under discussion — that the elephant attempted to catch the man first, in order to kill him afterwards. But prehensile organs are not used as such in the way described. That Andersson was about to be seized was purely suppositious upon his part, while the descent of the elephant's proboscis, with such violence that it swept away large stones as if they had been pebbles, was a matter of fact. The animal *did* strike, whether he intended to do so or not, and that this was not his intention is merely a guess. This story illustrates other traits also, and among these the alleged fear of man. "An implanted instinct of that kind," observes William J. Burchell ("Travels into the Interior of Southern Africa") "such as all wild beasts have, their timidity and submission, form part of that wise plan predetermined by the Deity, for giving supreme power to him who is, physically, the weakest of them all." The only objection to this very orthodox statement is that it is not true. Man is not weaker than many wild animals, and so far as "timidity and submission" go, he might have found African tribes barricading their villages and sleeping in trees for no other purpose than to keep out of their way. Caution proceeds from apprehension, and this from an experience of peril. When the conditions of existence are such that certain dangers persist, wariness in those directions originates and becomes hereditary. Man has been the elephant's constant foe, and in those places where human beings were able to destroy them, these animals were overawed; but otherwise not, or at least, certainly not in the sense in which this assertion is generally made. With regard to the conclusions — many of them directly contradictory — which prevail concerning the elephant's sense of smell, there are several circumstances which ought to be taken into consideration, but with the exception of currents of air, they have not been noticed to the author's knowledge. Scent in an elephant is very acute, and the scope of this sense, as well as its delicacy and discrimination, is greater than in most animals. At the same time, the nervous energy that vitalizes this apparatus is variable in quantity, and never exceeds a definite amount at any one time. If wind sweeps away those emanations which would otherwise have stimulated the olfactories, no result occurs, and precisely the same consequence follows a diversion of nerve force into other channels.

Many accounts have been given in which this seemed to be the cause of an unconsciousness that was explained by saying that the sense itself was in fault. Evidently, however, when the energy through which an organ acts is fully employed in carrying on action somewhere else, its function must be temporarily checked. Preoccupation, however, fully accounts for the phenomenon. Thought, feeling, concentrations of attention, physical and mental oscillations of many kinds, perturb, check, pervert, augment, or diminish function in this and other directions. If we cannot accustom ourselves to looking upon wild beasts as acting consciously and voluntarily, it seems probable that little progress towards understanding their habits and characters is likely to be made.

How, for example, are the following facts related by Gordon Cumming, to be reconciled with conventional opinions upon the shyness and timidity of elephants, their fear of man, and the possession of instincts which act independently of experience. It was in comparatively early times that these events took place, before many Europeans with rifles had gone into Africa, and when elephants knew less about firearms than they did when the big tusker nearly finished Andersson. "Three princely bulls," says Colonel Cumming, "came up one night to the fountain of La Bono." They knew that a man was there, for they had got his wind. It is possible that they also knew he was not a native, but if this were the case, that was all that they knew.

The leader was mortally wounded at about ten paces from the water, went off two hundred yards, "and there stood, evidently dying." His companions paused, "but soon one of them, the largest of the three, turned his head towards the fountain once more, and very slowly and warily came on." At this moment the wounded elephant "uttered the cry of death and fell heavily to the ground." The second one, still advancing, "examined with his trunk every yard of ground before he trod on it." Evidently there was no dancing, screaming horde of negroes with assegais about; equally sure was it that danger threatened from human devices, and the elephant, not being inspired as is commonly supposed, was looking for the only peril he knew anything about; that is to say, a pit-fall. As for the explosion and flash, these most probably were mistaken for thunder and lightning. In this manner, and with frequent pauses, this animal went round "three sides of the fountain, and then walked up to within six or seven yards of the muzzles of the guns." He was shot and disabled at the water's edge. By this time ignited wads from the pieces discharged had set fire to a bunch of stubble near by, and two more old bulls who followed the original band, went up to the blaze; one, the older and larger, appearing to be "much amused at it." This tusker staggered off with a mortal wound, and another came forward and stood still to drink within half pistol-shot of Colonel Cumming, who killed him. Three more male elephants now made their appearance, "first two, and then one," and of these two were shot, though only one of them fatally. What possible explanation can the doctrine of instinct give of such behavior as this upon the part of wild beasts? How does this kind of conduct accord with the idea of a ready-made mind that does not need to learn in order to know? In what manner shall we adjust such conduct to preconceptions concerning natural timidity and that implanted fear of man "predetermined by the Deity"? It may be said, of course, that Colonel Cumming's account was overdrawn; but the reply to an objection of this kind is that, overwhelming evidence to the same effect could be easily produced.

When an observant visitor walks along the line of platforms in an Indian elephant-stable, the differences exhibited by its occupants can scarcely fail to attract attention; and with every increase in his knowledge, these diversities accumulate in number and augment in importance. During the free intercourse of forest life, some influence, most probably sexual selection, has produced breeds whose characteristics are unmistakable. Even the uninitiated may at once recognize these. Koomeriah, Dwásala, and

Meèrga elephants exhibit marked contrasts, and experience has taught Europeans their respective values. The first is the best proportioned, bravest, and most tractable specimen of its kind; but it is rare. Intermediate between the thoroughbred and an ugly, "weedy," and in every way ill-conditioned Meèrga, comes what is called the Dwásala breed, to which about seventy per cent of all elephants in Asia belong. "Whole herds," says Sanderson, "frequently consist of Dwásalas, but never of Koomeriahs." Almost all animals used in hunting are of this middle class, and they constitute by far the largest division of those kept by the government. Females greatly outnumber males, and it may be owing to this fact that so many have been used in the pursuit of large game, although some famous sportsmen maintain that these are naturally more courageous and stancher than tuskers.

Great as are the unlikenesses seen among inmates of an establishment like that at Teperah, they will be found to be fully equalled by their dissimilarities in character; and those who have become familiar with elephants come to see that their dispositions and intelligence are to some extent displayed by their ordinary demeanor and looks. It is wonderful how much facial expression an elephant has. The face-skeleton is imperfect; that is to say, its nasal bones are rudimentary, while the mouth, and in fact all of the lower half of the face, is concealed beneath the great muscles attached to the base of the trunk. But in spite of that, and with his ears uncocked and his proboscis pendant, an elephant's countenance is full of character.

Passing along the lines where they stand, shackled by one foot to stone platforms, one sees, or learns to see, the individualities their visages reveal. Occasionally a heavily-fettered animal is met with, whose mien is disturbed and fierce. In his "little twinkling red eye," says Campbell, "gleams the fire of madness." He is "must"; the victim of a temporary delirium which seems to arise from keeping male elephants apart from their mates. But at length, amid all the appearances of sullenness, good nature, stupidity, bad temper, apathy, alertness, and intelligence, which the visitor will encounter, a creature is met with in whose ensemble there is an indescribable but unmistakable warning. Go to his keeper and state your views. That "true believer," if he happens to be a Mussulman, having salaamed in proportion to his expected bucksheesh, and said that Solomon was a fool in comparison with yourself, will then express his own sentiments but not so that the animal can hear him. These are to the effect that this elephant is an oppressor of the poor, a dog, a devil, an infidel, whose female relations to the remotest generations have been no better than they should be. That the kafir wants to kill him; is thinking about doing it at that moment, but *Ul-humd-ul-illa*, praise be to God, has not had a chance; though if it be his destiny, he will do so some day. Very probably these are not empty words. Most frequently the man knows what he is talking about. Still if one naturally asks, why then he stays in such a position, the answer breathes the very genius and spirit of the East. "Who can escape his destiny?" asks the idiotic fatalist, and remains where he is.

The systems of rewards and punishments by which discipline is kept up in a large elephant stable, affords several items of interest with respect to the character of these beasts. If, as sometimes is the case, an elephant shirks his work, or does it wrong on purpose, is mutinous, stubborn, or mischievous, a couple of his comrades are provided with a fathom or two of light chain with which they soundly thrash the delinquent, very much to his temporary improvement. This race is very fond of sweets, and sugar-cane or goor — unrefined sugar — forms an efficient bribe to good behavior. The animals take to drink very kindly, and when their accustomed ration of rum has been stopped for misconduct, they truly repent. Mostly, however, elephants are quiet, kindly beasts, and it is said by those who ought to know, that animosity is not apt to be cherished against men who correct them for faults of which they are themselves conscious. At the same time, nobody, if he is wise, gives an elephant cause to think himself injured. Very often the creature entertains this idea without cause, and it is not uncommon for them to conceive hatreds almost at first sight. D'Ewes ("Sporting in Both Hemispheres") relates one of the many reliable incidents illustrative of the animal's implacability when aggrieved. A friend of his, a field officer stationed at Jaulnah, owned an elephant remarkable for its "extreme docility." One of the attendants — "not his mahout" — ill-treated the creature in some way and was discharged in consequence. This man left the station; but six years after he, unfortunately for himself, returned, and walked up to renew his acquaintance with the abused brute, who let him approach without giving the least indication of anger, and as soon as he was close enough, trampled him to death. This is the kind of anecdote which Professor Robinson remarks is "infinitely discreditable to the elephant"; that fact, however, has nothing to do with the truth. All those good qualities the creature possesses can be done justice to without making any excursions into sentimental zoölogy. Captain A. W. Drayson ("Sporting Scenes in Southern Africa") asserts that "the elephant stands very high among the class of wild animals." That means nothing; affords no help to those who are trying to find out how high it stands. Sir Samuel Baker ("Wild Beasts and their Ways") gives his opinion more at length. Of the animal's sagacity he observes that it is, according to his ideas, "overrated. No elephant," he says, "that I ever saw, would spontaneously interfere to save his master from drowning or from attack. . . . An enemy might assassinate you at the feet of your favorite elephant, but he would never attempt to interfere in your defence; he would probably run away, or, if not, remain impassive, unless especially ordered or guided by his mahout. This is incontestible. . . . It is impossible for an ordinary bystander to comprehend the secret signs which are mutually understood by the elephant and his guide." Baker holds, with others who have really studied elephants, that when they evince any special sagacity, it is because they act under direction, and that if left to themselves they usually do the wrong thing. The species is naturally nervous, and this disability is increased by those alterations in its way of life that domestication involves. Captivity likewise shortens its existence. Profound physiological changes are thus produced, the most noticeable of which are barrenness, great capriciousness of appetite, enfeeblement of the digestive functions, and a marked vice of nutrition by which an animal that recovers from injuries the most severe in its wild state now finds every trifling hurt a serious matter, and often dies from accidents that would otherwise have been of little moment. In the same category must also be ranked the decreased endurance of tame elephants. The Asiatic species is much inferior to the African in this respect, by nature, but both sensibly deteriorate in this way when domesticated.

There is nothing to show that the African elephant is worse tempered than the Asiatic. It has never been reclaimed by the natives, and that fact no doubt has given rise to the opinion. In the Carthaginian, Numidian, and Roman provinces, this species was made use of very much as the other is now in India, and most if not all the famous homicidal elephants we know of, belonged to the latter country. But it would appear that a "rogue," properly so called, requires peculiar conditions under which to develop. "Rogue elephants," says Drummond, "are rare; indeed, it seems to me that it is necessary for the full formation of that amiable animal's character that it should inhabit a well-populated district where continual opportunities are afforded for attacking defenceless people, of breaking into their fields, and, in general, of losing its *natural respect for human beings;* and as such conditions seldom exist in Africa, from the elephant chiefly inhabiting districts devoid of population on account of their unhealthiness, the rogue, properly so called, is seldom met with, though the solitary bull, the same animal in an earlier stage, is common enough."

Drummond, it will be observed, clings to the superstition of man's recognized primacy in nature; and if he had declared that his appointment to this position was handed down by tradition among elephants from the time of Adam and the garden of Eden, the absurdity could scarcely be greater. In what possible way can a wild beast that has not been hunted know anything about a man, except that he is an unaccountable-looking little creature, who walks like a bird, and has a very singular odor?

A rogue who infested the Balaghat District is described by Baker as a captured elephant who after a considerable detention escaped to the forest again. "Domestication," he remarks, "seems to have sharpened its intellect and exaggerated its powers of mischief and cunning. . . . There was an actual love of homicide in this animal." He continually changed place, so that no one could foretell his whereabouts, and approached those whom he intended to destroy with such fatal skill that they never suspected his presence until it was too late. He made the public roads impassable. By day and night the inhabitants of villages lying far apart heard the screams which accompanied his attack, and immediately this monster was in the midst of them, killing men, women, and children. At length Colonel Bloomfield, aided by the whole population, succeeded in hunting the beast down. "Maddened by pursuit and wounds, he turned to charge," and as he lowered his trunk when closing, a heavy rifle ball struck him in the depression just above its base, and he fell dead.

Cunning as this elephant was, his actions displayed that lack of inventiveness which Sanderson charges against the race; and this defect saved the lives of many who would otherwise have been killed. If any one was out of reach in a small tree, the rogue never thought of getting at him by shaking its trunk. Both Sir Samuel and Captain R. N. G. Baker report having seen an elephant butt at a *Balanites Egyptiaca* when it was three feet in diameter, so that a man "must have held on exceedingly tight to avoid a fall." It is certain that these animals are accustomed to dislodge various edibles by this means. But a change in circumstances prevented the Balaghat brute from resorting to a well-known act which would have lengthened considerably the list of his victims.

Places in Africa where elephants once abounded now contain none. They are less subject to epidemics than many species, but suffer from climatic disorders and the attacks of parasites. This, however, is not the reason for their disappearance from certain localities. They have fallen before firearms, or migrated in fear of them. "From my own observation," says Baker, "I have concluded that wild animals of all kinds will withstand the dangers of traps, pit-falls, fire, and the usual methods employed for their destruction by savages, but will be speedily cleared out of an extensive district by firearms."

A field naturalist coming from Africa to India, or any other part of Asia, would be at once struck by the inferior size, darker color, smaller ears, less massive tusks (rudimentary in the female), and other structural differences presented by *Elephas Indicus*. Likewise, with the forest life, browsing habits, and nocturnal ways of this species, "there is little doubt that there is not an elephant ten feet high at the shoulder in India," says Sanderson. If a stranger took to elephant-hunting, his opinion of their character in that country would probably depend upon the escapes he made from being killed. There is, however, something yet to be said upon the subject of Asiatic rogues that, so far as the author is aware, has escaped the attention of those who have described them. Such creatures as those of Kakánkōta, Balaghat, Jubbulpūr, and the Begapore canal, are extremely exceptional, if what they actually did be alone considered, but there is nothing to show that they were very extraordinary in temper or traits of character. The first seems to have been undoubtedly insane; the others, however, gave no indications of mental alienation. They were simply vicious like great numbers of their kind, and the accidents of life enabled them to show it more conspicuously than is often the case. Whatever may be thought of the influence of descent in these instances, it is certain that a criminal class cannot develop itself among elephants, and that those murderous brutes referred to, do not stand alone.

Colonel Pollok ("Natural History Notes") gives a report extracted from the records in the Adjutant General's Office, that brings out several points relating to the character of vicious elephants. The statements made seem to be incredible, but those who have made a study of the subject will recall many examples of desperation, tenacity of life, and ferocity in elephants, that may serve to modify doubt ; more especially in connection with the effects of wounds in the head, which is so formed that half of it might be shot away without an animal suffering otherwise than from shock and loss of blood.

To C. SEALY, Magistrate, etc.

Sir : — I have the honor to state that on the 24th instant, at midnight, I received information that two elephants of very uncommon size had made their appearance within a few hundred yards of the cantonment and close to the village, the inhabitants of which were in the greatest alarm. I lost no time in despatching to the place all the public and private elephants we had in pursuit of them, and at daybreak on the 25th, was informed that their very superior size and apparent fierceness had rendered all attempts at their seizure unavailing; and that the most experienced mahout I had was dangerously hurt — the elephant he rode having been struck to the ground by one of the wild ones, which, with its companion, had then adjourned to a large sugar-cane field adjoining the village. I immediately ordered the guns (a section of a light battery) to this place, but wishing in the first place, to try every means for catching the animals, I assembled the inhabitants of the neighborhood, and with the assistance of the resident Rajah caused two deep pits to be prepared at the edge of the cane field in which our elephants and the people contrived, with the utmost dexterity, to retain the wild ones during the day. When these pits were reported ready, we repaired to the spot, and they were cleverly driven into them. But, unfortunately, one of the pits did not prove to be sufficiently deep, and the one who escaped from it, in the presence of many witnesses, assisted his companion out of the other pit with his trunk. Both were, however, with much exertion, brought back into the cane, and as no particular symptoms of vice or fierceness had appeared in the course of the day, I was anxious to make another effort to capture them. The beldars, therefore, were set to work to deepen the old and prepare new pits against daybreak, when I proposed to make the final attempt. About four o'clock yesterday, however, they burst through all my guards, and making for a village about three miles distant, reached it with such rapidity that the horsemen who galloped before them, had not time to apprise the inhabitants of their danger, and I regret to say that one poor man was torn limb from limb, a child trodden to death, and two women hurt. Their destruction now became absolutely necessary, and as they showed no disposition to quit the village where their mischief had been done, we had time to bring up the four-pound pieces of artillery [these events took place in 1809] from which they received several rounds, both ball and abundance of grape. The larger of the two was soon brought to the ground by a round shot in the head; but after remaining there about a quarter of an hour, apparently lifeless, he got up again as vigorous as ever, and the desperation of both at this period exceeds all description. They made repeated charges on the guns, and if it had not been for the uncommon bravery and steadiness of the artillerymen, who more than once turned them off with shots in the head and body when within a very few paces of them, many dreadful casualties must have occurred. We were obliged to desist for want of ammunition, and before a fresh supply could be obtained, the animals quitted the village, and though streaming with blood from a hundred wounds, proceeded with a rapidity I had no idea of towards Hazarebaugh. They were at length brought up by the horsemen and our elephants, within a short distance of a crowded bazaar, and ultimately, after many renewals of most formidable and ferocious attacks on the guns, gave up the contest with their lives.

The western half of those central Indian highlands called locally the Mykal, Máhádeo, and Sátpúra hills, is a famous haunt for elephants. In this wild birthplace of the streams that pour themselves into the Bay of Bengal and the Arabian Gulf, these creatures wander in comparative security. The Gónd, Kól, and Sántál aborigines furnish the best trackers extant, except, perhaps, those mysterious Bygá or Bhúmiá, whose knowledge of woodcraft is unequalled. These small, dark, silent men have no sort of respect for an elephant's mind or character, but they worship it from fear; they adore the animal because they know enough of its disposition to be always apprehensive of its doing more than it generally does.

Most of these great timber districts are under the supervision of officers, and the camps of their parties are widely scattered through large and lonely tracts of woodland. If one of these is come upon by a herd of elephants while its occupants are absent, a striking trait in this creature's character will almost surely be exhibited. No monkey is more mischievous than one of these big brutes, and when the men return they probably find that nothing which could be displaced, marred, or broken, has escaped their attention. Elephants are also very curious; anything unusual is apt to attract them, and if they do not become alarmed at it, the gravity with which a novel object is examined, and the queer, awkward way in which these beasts manifest interest or amusement, is singular enough. Sometimes their performances under the incitement of curiosity or malicious mischief are decidedly unpleasant. A wild elephant came out of the woods one night and pawed a hole in the side of Sanderson's tent. Hornaday says he made a little door in the wall at the head of his bed, so that he could bolt at once in case of a visitation like this. People living in such places, and in frail houses, are exposed to another contingency. Elephants are very subject to panics, and as they often arise from causes that should not disturb such a creature at all, no one can tell when a herd may not rush off together, and go screaming through the wood, breaking down everything but the big trees before them.

Sooner or later, a hunting party's progress will be arrested by the halt of their guide: he crouches down in his tracks and looks intently, as it appears, at nothing.

What he sees would be nothing to eyes less practised, but it is an elephant's spoor. If one were in Africa, the trackers would now smooth off a little spot of ground, make a few incantations, and throw magic dice to find out all about this animal. But here nothing of that kind is done, and yet the guide will follow the trail unerringly, and the hunter may count upon being brought to his game. "When you know," says Captain A. W. Drayson, "that the giant of the forest is not inferior in the senses of hearing and smell to any creature in creation, and has besides intelligence enough to know that you are his enemy, and also for what purpose you have come, it becomes a matter of great moment how, when, and where you approach him."

Elephants, unless they have some definite end in view, stroll about in the most desultory, and, if one is following them, the most exasperating manner. Their big round footprints go up hill and down dale in utterly aimless and devious meanderings. Here the brute stops to dig a tuber or break a branch, there for the purpose of tearing down a clump of bamboos, in another place with no object in view except to drive its tusks into a bank. Sportsmen often spend a day and night upon their trail.

No one can foresee the issue of a contest with an elephant. It may fall to a single shot, but no matter how brave and cool and well instructed the hunter may be, how stanch are his gun-bearers, how perfect his weapons and the skill with which they are used, when that wavering trunk becomes fixed in his direction, and the huge head turns toward him, his breath is in his nostrils. More than likely the animal, whose form is almost invisible in the half-lights of these forests, is aware of his pursuer's presence before the latter sees him, and if he has remained, it is because he means mischief. Then it may well happen with the sportsman as it did with Arlett, Wedderburn, Krieger, McLane, Wahlberg, and many another.

It stands to reason that a herd is harder to approach without being discovered than a single elephant would be. The chances that the hunter will be seen are greater, and their scattered positions make it more probable that some of them will get his wind.

Occasionally an old bull who despises that part of mankind who do not possess improved rifles, and knows perfectly well the difference between an Englishman and a native, will take possession of some unfortunate ryot's millet field or cane patch, and hold it by right of conquest against all attempts to dislodge him. Crowds revile the animal from a safe distance, and a village shikári comes with a small-bored matchlock and shoots pieces of old iron and pebbles at him from the nearest position where it is mathematically certain that he will be secure. As for the marauder, he stays where he is until everything is eaten or destroyed, or until he gets tired.

The amount actually consumed by elephants forms but a small portion of the loss which agriculturists sustain from their forays. They always trample down and ruin far more than they eat. Both in India and Ceylon, various districts suffered so severely in this way that government gave rewards for all elephants killed. This has now been discontinued in both countries, but in many places where the herds are protected their numbers are increasing, so that the same necessity for thinning them out will again arise.

All over the cultivated portions of India platforms are erected in fields, where children by day, and men at night, endeavor to frighten away these invaders, together with the birds, antelopes, bears, monkeys, and wild hogs, that ravage their crops. No very signal success can be said to attend these efforts, and when a herd of elephants makes its appearance, they simply keep at a distance from the stages, and otherwise do as they please.

Plundering bands survey the ground, study localities, go on their *duroras* like a troop of Dacoits, and are organized for the time being in a rude way, under the influence of what Professor Romanes calls "the collective instinct."

Hunters favorably situated can easily see this. A far-off trumpet now and then announces the herd's advance through the forest, but as they approach the point where possible danger is to be apprehended, no token of their presence is given, and its first indication is the appearance of a scout, — not a straggler who has got in front by accident, but an animal upon whom the others depend, and who is there to see that all is safe. Everything about the creature, its actions and attitudes, the way it steps, listens, and searches the air with slowly moving trunk, speaks for itself of wariness, knowledge of what might occur, and an appreciation of the position it occupies; no doubt, to a certain extent, of a sense of responsibility. When this scout feels satisfied that no danger is impending, it moves on, at the same time assuring those who yet remain hidden that they may follow, by one of the many significant sounds that elephants make.

A number of narratives describe events as they are likely then to occur, but they are merely hunting stories, and so far as the writer's memory serves, do not bring out the animal's traits in any special way. It would appear, however, that the behavior of elephants who unexpectedly meet with Europeans in those places where all the resistance previously experienced came from farmers themselves, is very different from what it is in the former case. Then they are said to be difficult to get rid of, and when driven away from one point by shouts, horns, drums, and the firing of guns, they rush away to another part of the plantations, and continue their depredations. No such passive resistance as this is attempted when English sportsmen are upon the spot. Elephants discover their presence immediately. Upon the first explosion of a heavy rifle, the alarm is sounded from different parts of the field, and the herd betakes itself to flight without any notion of halting by the way. Their dominant idea is to get clear of those premises as soon as possible.

"The elephant," says Andersson, "has a very expressive organ of voice. The sounds which he utters have been distinguished by his Asiatic keepers into three kinds. The first is very shrill, and is produced by blowing through his trunk. This is indicative of pleasure. The second, made by the mouth, is a low note expressive of want; and the third, proceeding from the throat, is a terrific roar of anger or revenge." Sanderson seems to think that these discriminations are somewhat fanciful. He remarks that "elephants make use of a great variety of sounds in communicating with one another, and in expressing their wants and feelings." But he adds that, while "some are made by the trunk and some by the throat, the conjunctures in which either means of expression is employed, cannot be strictly classified, as pleasure, fear, want, and other emotions are indicated by either." Leveson, on the contrary, gives a list of these intonations, and describes the manner in which they are produced. So also does Tennant; and Baker adds another sound to those before given; "a growl," this writer calls it, and he says that "it is exactly like the rumbling of distant thunder."

Undoubtedly these animals express their thoughts and feelings intelligibly by the voice, as also through facial expressions, and by means of such gestures as they are capable of making. It has been before said that although the elephant's face is half covered up, and there are no muscles either in his case or in that of any other animal, whose primary function is to express mental or emotional states, his physiognomy may be in the highest degree significant.

"The courage of elephants," writes Captain Drayson, "seems to fluctuate in a greater degree than that of man. Sometimes a herd is unapproachable from savageness; sometimes the animals are the greatest curs in creation." What is called boldness varies considerably in different species, among members of the same species, and in the same individuals at different times. It is a quality, that, like all others, is double-sided, certain elements belonging to the mind, and the residue to the body. Elephants are nervous; that is to say, their nerve centres — the ganglia in which energy is stored up — are constitutionally in a state of more or less unstable equilibrium, so that stimulus, whether of external origin, or initiated centrically, is apt to produce explosive effects. Courage depends upon physical and mental constitution, upon specializations in race, training, and structure, upon differences in personal experience and organization.

So much as this may be said with confidence, but on what grounds, biological or psychological, is it possible for Professor Romanes to assert that the elephant seems usually to be "actuated by the most magnanimous of feelings"? Magnanimity belongs to the rarest and loftiest type of human character: how did an elephant come by it? The obligations of mental and moral congruity are not less binding than those of physical fitness. No one nowadays draws an elephant with a human head; but a beast with self-respect, courage, refinement, sympathy, and charity enough to be magnanimous, does not seem to outrage any sense of propriety. Works like those of Watson ("Reasoning Power of Animals"), Broderip ("Zoölogical Recreations"), Bingley ("Animal Biography"), Swainson ("Habits and Instincts of Animals"), too often interpret facts so that they will fit preconceived opinions. There is a story, for example, by Captain Shipp, of how, during the siege of Bhurtpore, an elephant pushed another one into a well because he had appropriated his bucket. Tales like this resemble pictures in which the design and execution are both weak, and which depend for their effect upon accessories illegitimately introduced into the composition. Probably a large part of the present inhabitants of the earth have seen animals who, while contending for some possession, acted in a similar manner; but they were not elephants, nor were the circumstances of a well and a siege at hand to set them off, and produce an impression that the actual incident does not justify. The grief of captive elephants over their situation is a subject upon which many fine remarks have been made. Colonel Yule ("Embassy to Ava") states that numbers die from this cause alone; but *yaarba'hd*, either in its dropsical or atrophic form, is what chiefly proves fatal to them, and this is brought on by the sudden and violent interruption of their natural way of life. According to Strachan, Sanderson, and other experts, the disorder is due to an overthrow of functional balance; something which is sure to induce disease whenever it occurs. Sterility, temporary failure of milk in females with calves, together with the various effects already mentioned, may be referred to the same cause. It is not said that elephants never die of grief; still less, that this is impossible. Any animal highly organized enough to feel intense and persistent sorrow may perish. Pain, either physical or mental, is intimately connected with waste of tissue and paralysis of reparative action. Bain's formula that "states of pleasure are concomitant with an increase, and states of pain with a decrease, of some, or all, of the vital functions," is not strictly correct as it stands; still the truth it is intended to convey remains indisputable. Grant-Allen ("Physiological Æsthetics") defines pleasure as a "concomitant of the healthy action of any or all of the organs or members supplied with afferent cerebro-spinal nerves, to an extent not exceeding the ordinary powers of reparation possessed by the system." Grief, when intense, reverses this, makes normal function impossible, palsies the viscera, and impairs or perverts those nutritive processes upon which life directly depends. But the profound and abiding sorrow this race cherishes in servitude is a romance. There is nothing to show the regret and longing which have been imagined. Elephants struggle for a while against coercion, and then forget. They fail to take advantage of opportunities for escape, and when they do, the fugitives are recaptured more easily than they were taken in the first place. Instances have often occurred of their voluntary return after a long absence. In the beginning, it is the finest animals who perish. They kill themselves in their struggles, or die of disease. Subsequently, it is said that domestication lengthens average life. This must, however, be one of those blank assertions made so commonly about wild beasts; since, independently of any other objection, it is evident that the statement, in order to be worth anything, should rest upon the basis of a wide comparison between the relative longevities of free and captive animals, and vital statistics of this kind, not only have not been tabulated, but it is impossible that they should have been collected.

Colonel Pollok remarks that "at all times, this is a wandering race, and consumes so much, and wastes so much, that no single forest could long support a large number of such occupants." Livingstone, Forsyth, and others have, however, noted the fact that little or no permanent injury to extensive woodlands was wrought by these animals. They do not overturn trees, as is popularly believed, and still less do they uproot them. Elephants bend down stems by pressure with their foreheads, and they go loitering about breaking branches, till the place looks as if a whirlwind had passed over it, but these devastations are of a kind soon repaired. In the forests of India they have never met with such adversaries, or been exposed to the same dangers, as the species encountered on the "Dark Continent." Some Indian tribes worshipped, and all feared them. They passed their lives for the most part in peace, finding food plentiful, ruining much, and finishing nothing. Pitfalls were few and far between; no weighted darts fell upon them as they passed beneath the boughs, no pigmy savage stole behind as they leaned against a tree boll and woke the echoes of the wood with deep, slow-drawn, and far-resounding snores, to thrust a broad-bladed spear into their bodies, and leave it there to lacerate and kill his victim slowly. Neither were herds driven over precipices, nor into chasms, nor did hordes of capering barbarians come against them with assagais, and scream, while pricking them to death, —

"Oh Chief! Chief! we have come to kill you,
Oh Chief! Chief! many more shall die.
The gods have said it."

All this was common throughout Africa, while in Asia the natives seldom aggressed against elephants except in the way of capturing them. It is true that this was done awkwardly, and often caused injury or death; but that was unintentional, and as a rule they roamed unmolested among the solitudes of nature.

Existence had its drawbacks, however. Elephants were not eaten in Asia, and not hunted for their ivory to any extent, but they were used in war, and the state of no native prince could be complete unless he had an elephant to ride on and several caparisoned animals for show. Owing to these needs and fashions the animals were captured extensively. In many places at present small parties of men, often only two or three, go on foot into the forests as their predecessors did ages ago, each with a small bag of provisions, and a green hide rope capable of being considerably stretched. An elephant's track is almost as explicit and full of information to them as a passport or descriptive list, and when they have found the right one, it is patiently followed till the beast that made it is discovered. Then in the great majority of cases its fate is fixed. Flight, concealment, resistance, are in vain. In some "inevitable hour" a noose of plaited thongs that cannot be broken is slipped around one of the hind feet, and a turn or two quickly taken about a tree. A high-bred elephant gives up when he finds that the first fierce struggle for freedom is unavailing, but the meerga's resistance lasts longer. After one leg has been secured it is easy to fetter both, and then the captors camp in front of the animal in order to accustom it to their presence. By degrees they loosen its bonds, feed and pacify it. When anger is over, and its terrors are dissipated, these men lead their captive off to a market at some great fair, and they lie about what they have done and what the elephant did, with a fertility of invention, a height and length and breadth of mendacity which it would be vain to expect to find exceeded in this imperfect state of existence.

The government also often wants elephants, and when this is the case, captures are made in a different manner, and upon a greater scale. What is done is to surround a herd and drive it into an enclosure called a keddah. This is often a very complicated and difficult thing to accomplish. Far away in some wild unsettled region of the Nilgiri or Satpúra hills, the uplands of Mysore, or elsewhere, an English official pitches his tent, surveys the country, and sends out scouts. To him sooner or later comes a person without any clothes to speak of, but with the most exquisite manners, and says that, owing to his Excellency's good fortune, by which all adverse influences have been happily averted, he begs to represent that a herd of elephants, who were created on purpose to be captured by him, is marked down. Then the commander-in-chief of the catching forces opens a campaign that may last for weeks, or even months. The topography has been carefully studied with reference to occupying positions which will prevent the animals from breaking through a line of posts that are established around them, and between which communication is kept up by flying detachments. Drafts of men from the district and a trained contingent the officer brought with him, are manœuvred so that they can concentrate upon the point selected for their keddah, which is not constructed till towards the close of these movements, since the area surrounded is very extensive and it is not at first known exactly where it must be placed. Its position is fixed within certain limits, however, and their object is to drive the herd in that direction without at first attracting attention to the fact that this is being done, and thereby causing continued alarm. Those who direct proceedings know the character of elephants, and count upon their lack of intelligence to aid them in carrying out the design. Before any apprehension of real danger makes itself felt, they have voluntarily, as it seems to them, moved away from parties who just showed themselves from time to time and then disappeared. They still feed in solitudes apparently uninvaded, still stand about after the manner of their kind, blowing dust through their trunks or squirting water over their bodies. They fan themselves with branches, and sleep in peace.

At length, long after the true state of things would have been fully appreciated by most other species, the herd finds out that it is always moving in a definite direction. Then a dim consciousness of the truth, which day by day becomes more vivid until it arrives at certainty, takes possession of their minds. From that time an exhibition of traits which scarcely correspond with popular views upon the elephant's intellect is constantly made. If they had anything like the ability attributed to them, the toils by which they are surrounded could be broken with ease. There is no time from their first sight of a human being to the very moment when they are bound to trees, at which they could not escape. It is useless to say they do not know this; that is precisely what the creatures are accused of. If they were such animals as they are said to be, they would know it, and act accordingly. But as soon as the situation is revealed, they become helpless; their resources of every kind are at an end. They stand still in stupid despair, break out in transient and impotent fits of rage, make pitiable demonstrations of attack upon points where they could not be opposed for an instant if the assault was made in earnest, and at length suffer themselves to be driven into an enclosure that would no more hold them against their will than if it had been made of gauze.

An elephant corral or keddah is a stout stockade with a shallow ditch dug around it inside, and slight fences of brush diverging for some distance from its entrance. Incredible as it may seem, single elephants frequently break out of these places, but a herd hardly ever; they have not enterprise, pluck, and presence of mind enough to follow the example when it is set them. Sometimes, as we have seen, elephants may be fierce and determined; desperation has been shown to be among the possibilities of their nature. But whereas an exceptional individual will, from pure ferocity, brave wounds and death, nothing can so move the race as to cause a display of ordinary self-possession. It is quite true that whenever the imprisoned band comes rushing down upon any part of the keddah, they are met with fire-brands, the discharge of unshotted guns, and an infernal clamor; but if that be urged in explanation of their hesitation, it may be replied that if the whole herd had as much resolution as a single lion brought to bay, they would sweep away everything before them as the fallen leaves of their forests are swept away by a gale.

Often among the bewildered and panic-stricken crowd within a corral some animal is so dangerous that it has to be shot; the majority, however, soon grow calmer, and then comes the task of securing those which it is desirable to keep. When these are males, the procedure is as follows: An experienced female is introduced; she marches up to the tusker, and very shortly all sense of his situation vanishes from his "half-human mind." The fascinating creature who is made to cajole him has a man on her neck whose voice and motions direct her in everything she does; but that circumstance, which might undoubtedly be supposed to attract the captive's attention, is entirely overlooked, and when, either by herself or with the assistance of another Delilah, she has backed her Samson up against a tree, two or three other men who have been riding on her back, but whom he has not noticed, slip down and make him fast. As has been said, after a few fits of hysterics, his resistance is at an end; the monarch of the forest is tamed, and considering what has been written about elephants, it is indeed surprising that no one has reported the precise course of thought that produced his resignation. To express this change in the felicitous language of Professor Romanes, the elephant has experienced "a transformation of emotional psychology." That is to say, a being which has heretofore been nothing but an unreclaimed wild beast, is by the simple process of being frightened, deceived, abused, and enslaved, at once converted into one of the chief ornaments of animated nature!

The question arises as one ponders upon statements

like this, whether we really know anything worth speaking of about inferior animals, and if it is possible to use expressions like "cruel as a tiger," "brave as a lion," or "sagacious as an elephant," rationally. As for any philosophical, or, as Spencer calls it, "completely unified knowledge" on the subject, nobody possesses it; at the same time the natural sciences may be so applied as to bring certain truths to light in this connection. It is plain, for example, that an elephant does not kill his keeper because he is fond of him; but it is one thing to start out with the assumption that this noble-hearted, affectionate, and magnanimous animal would never have been guilty of such an act unless it had been maltreated, and it is another, and quite a different course to begin with the fact that the deed was done by a brute in whose inherited nature no radical change could by any possibility have been effected by such training as it has received. If now we endeavor to ascertain what that nature was, — study the records of behavior in wild and domesticated specimens, and look at this by the light which biology and psychology, without any assumptions whatever, cast upon it, — we shall find ourselves in the best position for investigating any particular case under consideration. Many accounts of such murders have been given at length. We know how, why, when, and where the animal began its enmity, and the manner in which it was shown or concealed, so that, having investigated the matter in the way described, we are, to a certain extent, able, not to generalize the character of this species, but to put aside immature opinions, and say that since very many elephants exhibit traits which are in conformity with those to be expected of them, these probably belong to the species at large, and may be displayed with different degrees of violence whenever circumstances favor their manifestation.

The chief characteristics of elephants have been discussed, and an attempt has been made to place them in their true light. The writer has not found the half-human elephant in nature, nor does it appear from records that any one else has done so. An elephant is a wild beast, comparatively with others undeveloped by a severe struggle for existence; superficially changed in captivity, and cut off from improvement by barrenness. It is capable of receiving a considerable amount of instruction, and learns quickly and well; but how far its acquisitions are assimilated and converted into faculty, is altogether uncertain. In the savage state elephants do nothing that other animals cannot do as well, and many of them better. Mere bulk, and its accompaniment, strength, do not influence character in any definite manner that can be pointed out.

In captivity, elephants are commonly obedient, partly because, having never had any enemies to contend with, they are naturally inoffensive, and partly for the reason that these animals are easily overawed, very nervous, and extremely liable to feelings of causeless apprehension.

Courage in cold blood is certainly not one of their qualities; nevertheless, being amenable to discipline, and having some sense of responsibility, certain elephants are undoubtedly stanch both in war and the chase.

This animal is easily excited, very irritable, prone to take offence, and subject to fits of hysterical passion.

Thus it happens that wild elephants are the most formidable objects of pursuit known to exist, and that the majority of those held in durance exhibit dangerous outbreaks of temper. When an elephant is vicious, he displays capabilities in the way of evil such as none of his kind, when left to themselves, have ever been known to manifest in the direction of virtue. A "rogue" is the most terrible of wild beasts; the captive tusker who has determined upon murder finds no being but man, who in the prosecution of his design is so patient, so self-contained, so deceitful, and so deadly. It is idle to say, speaking of the relations between elephants and men, that the good qualities of the former greatly predominate, since if it had been otherwise, no association between them would have been possible — they could not have inhabited the same regions.

The concluding pages may, perhaps, serve to show how far this sketch of the elephant's character is compatible with facts.

Charles John Andersson ("The Lion and the Elephant") observes that, "whether or not the elephant be the harmless creature he is represented by many, certain it is that to the sportsman he is the most formidable of all those beasts, the lion not excepted, that roam the African wilds; and few there are who make the pursuit of him a profession, that do not, sooner or later, come to grief of some kind." Being social animals, there is a certain sympathy and affection between members of the same family; but while striking instances of this are recorded, the bulk of evidence tends the other way.

Impressive examples of solicitude have, however, been observed. Moodie tells that he saw a female — whom the experience of most hunters shows to be much more likely to act in this manner than a male — guard her wounded mate, and how she, "regardless of her own danger, quitted her shelter in the woods, rushed out to his assistance, walked round and round him, chased away the assailants, and returning to his side caressed him. Whenever he attempted to walk, she placed her flank or her shoulder to his wounded side and supported him." Frederick Green wrote an altogether unique account to Andersson of the succor of an elephant that had been shot, by one who was a stranger, of the same sex, and who encountered him far from the scene where his misfortune had befallen him.

The Bushmen, he says, often asserted that elephants would carry water in their trunks to a wounded companion at a long distance in the "Weldt." Green, however, did not believe it, until, while hunting in the Lake Regions, he was compelled, from want of ammunition, "to leave an elephant that was crippled (one of his fore legs had been broken, besides having eleven wounds in his body) some thirty miles from the waggons."

"As I felt confident," this writer continues, "that he would die of his wounds . . . I despatched Bushmen after him instead of going myself; but they, not attending to my commands, remained for two days beside an elephant previously killed by my after-rider. It was, therefore, not until the fourth evening after I left this elephant that the Bushmen came up with him. . . . They found him still alive and standing, but unable to walk. . . . They slept near him, thinking he might die during the night; but at an early hour after dark they heard another elephant at a distance, apparently calling, and he was answered by the wounded one. The calls and answers continued until the stranger came up, and they saw him giving the hurt one water, after which he assisted in taking his maimed companion away." Such was the story told Green when the party came back. He disbelieved their statements entirely, went off to the spot to see what had happened for himself, and thus relates his own observations:—

"The next afternoon found me at the identical place where I had left the wounded elephant. I can only say that the account of the Bushmen as to the stranger elephant coming up to the maimed one was proved by the spoor; and that their further assertion as to his having assisted his unfortunate friend in removing elsewhere was also fully verified from the spoor of the two being close alongside of each other — the broken leg of the wounded one leaving after it a deep furrow in the sand. As I was satisfied that these parts of their story were correct, I did not see any further reason to doubt the other."

Male elephants rarely fall in the holes which undermine so many parts of Africa; they carry their trunks low, have no one to look out for but themselves, and so detect these traps, and generally uncover them. Moodie makes the statement that many elephants follow the recent trails of those who went before them to watering-places, and if these turned off, took it for a sign of danger, and did not drink. After what Inglis and Hallet say to the same effect of tigers, after St. John's observations upon red deer, and Lloyd's on the Scandinavian fox, inductive reasoning like this does not seem at all incredible. Amral, chief of the Namaqua Hottentots, told Galton and Andersson that on one occasion he and others were in pursuit of a herd of elephants, and at length came to a wagon-track which the animals had crossed. Here the latter, as was seen by their spoor, had come to a halt, and after carefully examining the ground with their trunks, formed a circle in the centre of which their leader took up his position. Afterwards individuals were sent out to make further investigations. The *Raad*, or debate, this chieftain went on to say, must have been long and weighty, for they (the elephants) had written much on the ground with their probosces. The decision evidently was that to remain longer in that locality would be dangerous, and they therefore came to the unanimous resolution to decamp forthwith. Attempts to overtake them, Amral went on to say, were useless; for, though they followed their tracks till sunset, they saw no more of them.

What these elephants thought when they found a track which, to them, was new and inexplicable, is, of course, a matter of conjecture; but their trail revealed everything that was done on this occasion, as clearly as if the Hottentots had been eye-witnesses of their actions.

Colonel Julius Barras ("India and Tiger-Hunting") entered *con amore* into a study of the elephant, so far as its character came into play when the animal was employed in sport; and he did what no other gentleman, to the author's knowledge, has ever done ; namely, turned mahout himself, and drove shikar tuskers against many a tiger. His appreciation of this creature's courage, benevolence, and reliability is very much in accord with that which has been expressed; but he offers some observations upon vice that should not be overlooked. "One peculiarity of elephants," remarks the Colonel, "is that, when desirous of killing any one, they nearly always select as a victim their own or a rival's attendant." It seems rather strained, however, to speak of this fact as a "peculiarity," since circumstances would naturally bring about such a selection.

But no provocation need be offered to an elephant in order that he should desire to kill a man. "Sahib," said Mohammed Yakoob, the driver of an immense old tusker, whom Colonel Barras had drawn from the government stables at Baroda, "you see that this elephant is a beast void of religion (*be imān*), and he hates the English."

"Dear me," answered the Colonel, "and how does he get on with the natives?"

"Oh!" replied the mahout, "much better, but still he is uncertain even with them. He has killed two, and there is but little doubt that he will do for me, his keeper, sooner or later."

Colonel Barras knew that Futtch Ali, the elephant in question, had never seen him before, and was well aware that it was impossible for this creature to feel offended at any act of his. The colonel's mind was also full of conventional ideas concerning elephants, so he disbelieved what the driver told him, and resolved to make friends with Futtch Ali, and ride him after tigers. He tells what happened in the following words :—

"One afternoon I considered myself fortunate in arriving before Futtch Ali when no one was in sight. I drew up in front of him with a few pieces of chopped sugar-cane in my hand. I looked attentively at the colossus, and could observe no signs of any unusual emotion. I spoke to him in those tones which I flattered myself he considered dulcet. On this he gently waved his ears and twinkled his eyes, as who should say, 'It's all right; you are my friend.' I now called out cheerfully, ' Salaam, Futtch Ali, Salaam!' and raised my arm at the same time. To this he responded by lifting his trunk over his head in return for the salute. This last act made assurance doubly sure. I mounted the platform, and as I did so the elephant again flung up his trunk, and opened his mouth, as if to accept with gratitude my sweet and juicy offerings. But his heart was full of treachery. He well knew that with his front feet manacled it would be useless to pursue me even if I had but a few inches start of him. He therefore dissembled with great cleverness and self-command till I had actually leant up against one of his tusks, and had got my hand in his mouth; then he suddenly belched forth a shout of rage, and made a sweep at me with his tusks that sent me flying off the platform into the dust below. . . . I sat up bareheaded and half-stunned, just in time to see the under-keeper, who had been slumbering behind a pile of equipments all this time, sent with greater force in a backward direction. . . . The elephant, meanwhile, had thrown off the mask; it was evidently only the shackles on his front feet that prevented him from getting off the platform and finishing us."

Very few persons would have done the same, but Colonel Barras took Futtch Ali for his Shikar elephant, and he afterwards carried him well in many a dangerous strait. But he was wise enough never to give him a second opportunity to take his life.

Another tusker enraged himself against Colonel Barras for a very slight cause. He was coming back one day, riding this animal, Ashmut Gūj by name, when, as he says, " I determined to see what this beast would do, if I, seated on his back, were to imitate a tiger charging." Accordingly, he began to mimic that short, hoarse, savage cry, and the elephant, who was not at all deceived, did nothing but raise his trunk. The mahout, however, warned him to desist. "Every time you make that noise," said he, "the elephant points his trunk over his back and takes a long sniff to inform himself as to which of his passengers is trying to vex him." Barras stopped at once, but the evil had been done.

"On arriving at the bungalow," the Colonel continues, " I had quite forgotten this little incident. Not so Ashmut Gūj. At the word of command he bent his hind legs and allowed the three natives to slip off his back in succession. I was the last to dismount, and as I touched the ground the elephant rose with a swift motion, and aimed a fearful kick at me with his enormous club-like hind foot. I started forward, so as just to escape the blow, which would, of course, have annihilated me. This elephant would never forgive me for the indignity I had put upon him. Always upon dismounting he would try to rise, so as to repeat his manœuvre, and it was necessary to make him kneel down completely before I got off. Nor would I ever again feed him from my hand, as I believe that if he could have got hold of me he would have trampled me."

There is a tragic story told by the same author, of an elephant who was "must." His keeper did not know it, and, in fact, could not be persuaded that such was the case.

Barras left Neemuch with a number of elephants, and among the rest an old friend and favorite of his, Roghanath Gūj, whose mahout, Ghassee Ram, had been in charge of him for eighteen years and thus acquired a very great influence over the animal. Colonel Barras, who had not seen this beast for some time, was at once struck by the indifference displayed to his expressions of friendliness, and to those little presents of sweets which these creatures enjoy so much. Evidently Roghanath Gūj was changed; ill, perhaps? No, said and swore his keeper, there was nothing the matter. His dulness, that sombre air which excited surprise and suspicion, was nothing more than a little irritability caused by the extremely hot weather. So Barras yielded his better judgment to greater experience, and the consequence was that the next day, while beating for a tiger, the elephant suddenly rushed upon one of the attendants, and would have killed him if the man had not taken off his turban and left it on a bush, while he himself slipped down into the shade of a deep ravine.

From this time forth Roghanath Gūj was picketed by himself.

"Two days after," says Barras, "we arrived at a small village,"—Mehra,—"and close to it there were some enormous Banyan trees, under which the elephants were secured. Opposite to them, on the other side of a small clearing, stood our little camp. Here, after a long and unsuccessful day's beating after a wary tiger, we enjoyed our late dinner, and had just sought our couches, clad for the night in our light sleeping-suits, when a burst of affrighted cries broke upon our ears. The tumult proceeded from the direction of the great tree where Roghanath Gūj stood in solitude.

"We instantly rushed for our guns, and seized a hurricane lamp. We made all haste in our slippered feet to the scene of action. As we got within twenty yards of the elephant, Ghassee Ram (his driver) called to us to halt. The animal, he said, was obeying him, and if nothing further incensed him, he would be able to tie up his hind legs with a rope, when he would be incapable (the fore-limbs being already chained) of doing any more mischief. So we stood where we were, and waited in great anxiety, whilst we could hear the mahout uttering the word *Sōm-Sōm*, which is the order for an elephant to keep his hind quarters towards any one who may be washing, or otherwise attending to him. The night was as dark as pitch; nothing could be seen. According to the different cries of the excited people, however, it was clear that something had happened to the under-keeper of Roghanath Gūj. Some said he was dead, some that he had escaped from his terrible assailant. I called to the other elephant-keepers, but they had all gone with their animals, I knew not whither, on the first alarm.

"Meanwhile Ghassee Ram was left quite alone to deal with the enraged beast. Of course we talked to him all the time, and were prepared to rush in and fire, as well as we could, if he called upon us to do so. Every chance, however, would have been against our disabling the elephant, who, maddened by such wounds as he might have received, would have worked untold destruction during the long dark hours of a moonless night. To the pluck of Ghassee Ram must be ascribed the avoidance of such a calamity. In a few minutes, which seemed an age, the mahout called out that we might advance. We did so, and never shall I forget the weirdness of the scene that was lighted up by the bright rays of the lamp I carried.

"Under the tree, and with his back to its stem, towered the dark form of the elephant, whilst his mahout, a mere speck, stood a little to his right. No other living being was visible, but close to the animal, on the opposite side from Ghassee Ram, lay a small, shapeless object, which a second glance showed to be the missing man. The elephant, with his ears raised, seemed to be keeping guard over his victim, and would probably kill any one who should attempt to remove the body, which lay within reach of his trunk. Still, this must be done, and at once, for life might yet be lingering in the shattered frame. I therefore gave the hurricane lamp to the mahout, and ordered him to swing it up in the elephant's face, and call out his name at the same time. Ghassee Ram, from the long habit of commanding this huge animal, had acquired some powerful tones. As he swung the lamp, that hung by a large ring, in the elephant's face, and cried out 'Roghanath Gūj, Roghanath Gūj,' the animal seemed deeply impressed. As the light ascended for the third time towards his dazzled eyes, I darted from between my two friends, who stood covering the elephant with their guns, and drew forth the unfortunate keeper. He was terribly mangled, and quite dead." This elephant was semi-delirious, and in that state the wild beast nature, which had been covered by a thin layer of educational polish, came out under the stimulus of some passing irritation. His mahout saw the man struck down, and interfered; but the animal was only restrained by his voice for a moment, and then completed the murder. He was not wholly demented, however; for Colonel Barras says, "I could not but be touched by the affection this huge creature displayed, even in his madness, towards the only two people he loved, — Ghassee Ram and myself. I fed him every day from my hand, and he never failed to clank his heavy chains, and turn round to watch me till I disappeared in my tent on leaving him."

It is probable that many persons whose minds are made up on the subject of elephants, may see nothing in this account but a case of perversion due to disease, and will pass by the elephant's evident power of self-restraint and discrimination as of no significance; contending that Roghanath Gūj, like all his kind, was naturally benevolent and amiable. Likewise, that the vagaries belonging to certain forms of mental alienation, temporary and chronic, are of the most eccentric and various character, and that this instance proves nothing with regard to the elephant's inherent nature. As a mere matter of reasoning, the objection is valid, and logically it is unanswerable; but, perhaps, some of those who believe that these brutes possess virtues of which most men are nearly destitute, will inform the world why "must"-delirium or actual insanity in an elephant, always takes the form of homicidal mania.

THE LION

"FROM the earliest times," says the writer on this subject in the "Encyclopædia Britannica," "few animals have been better known to man than the lion." It is precisely because of this knowledge, for the most part purely imaginary, that the real lion is less known than almost any of the other great wild beasts. Not so much in this case on account of the paucity of facts as from a plethora of fiction, his actual character has very imperfectly come to light.

Since Aristotle there have always been naturalists who contended for two species of these animals, and sometimes more.

In Greece, classification was made on the basis of size; in Rome, upon that of color. With regard to the first, Sir Samuel Baker remarks that the lions of Cutch and Guzrat are perhaps not so large as their African congeners; but according to Dr. Jerdon ("Mammals of India") measurements show that they are fully equal in this respect. Gérard, Livingstone, and others notice very discernible local contrasts in bulk among them in different parts of Africa itself, and it has been maintained by many that the lion grows smaller as one goes south from the Atlas. Major Smee has also been largely followed in his opinion

that the Asiatic, or more particularly the Indian, lion is maneless. Dr. Blyth, however, was able to demonstrate from the specimens in the Calcutta Museum that this was not the case, and his view of the accidental character of this deficiency is no doubt the true one. Frederick Courteney Selous ("A Hunter's Wanderings in Africa") paid particular attention to this feature, and he states that "out of fifty male lion skins, scarcely two will be found alike in the color or length of mane"; he adds that, judging from the same facts which those who multiply natural groups rely upon, "it would be as reasonable to suppose that there are twenty species as two."

This is but a hint at those discrepancies which have arisen from attaching different values to external and secondary characteristics. Antagonisms of this kind are overabundant, still there is no doubt that wherever lions now exist, they are specifically the same. There is but one genus of lion, with a single species, whose members vary in size, skin-appendages, color, temper, and habits, with the physiography of those provinces they inhabit, and of their human population, with breed, age, temperament, special environment, and their personal experience of men and things.

Sir Samuel Baker ("The Rifle and Hound in Ceylon") remarks in the course of his observations upon the Cingalese buffalo that no individual opinion upon the traits and disposition of an animal "can be depended upon," unless its pursuit "has been followed as a sport by itself." The results of many hunters' experiences are, however, on record, and so far as facts go, we are actually possessed of a more varied and extensive acquaintance with the species than any individual contact with it would be likely to give.

There is much that is inadequate and also illusory in Gérard's descriptions. Still, he met the formidable adversaries he encountered in a heroic spirit, and had seen them face to face too often not to be disabused of many errors. The sultan of the desert as known by him did not fear man, was not abashed in his presence, and could not be quelled by his eye. On the contrary, an attempt to stare him out of countenance was, as Sir Samuel Baker observes, the surest means to provoke an attack. Gérard's experience carried him too far. He only knew the lions of Algeria and Oran, but he thought that these animals were the same everywhere. Such is not the case. The race is now extinct in great areas where it was once distributed. No trace of it is left in many countries of Asia Minor, and it is dying out in Western Asia and India. In some regions man has exterminated the lion or driven him away, and there are other districts where this animal has learned that the battle nearly always goes against him, and where he now has to be forced to fight. On the other hand, certain tribes cower before lions, and this does not fail to change the relations they sustain towards mankind.

This imposing animal makes its appearance in art and literature very early. Frequent mention is made of it in the Cuneiform tablets and Hebrew Scriptures. In Pentaur's Egyptian Epic upon the War of Rameses II. against the Cheta or Hittites, lions are said to have accompanied the king's chariot, and fought as the Greek mastiffs (the dogs of Molossos) did at Marathon, or those of the British during Cæsar's invasion. Herodotus ("Polymnia") states that when Xerxes' hordes were moving in the country that lay between the rivers Nestos in Abedra, and Achelous of Acarnania, the camel trains suffered much loss from the attacks of these animals. He informs us that their range was restricted to this district, and expresses his surprise that camels, being creatures that these lions had never seen and might have been supposed to shun, were their especial victims. After Herodotus, when the Greeks began to write about everything that attracted their attention, much was said in one way or another concerning lions, but it amounted to no more than the little that can be found in Roman archives. It really seems as if classic writers left out on purpose everything that one would have cared most to know. Not even the minute and laborious scholarship of the sixteenth century, devoted as it mainly was to the explication of antiquity, has succeeded in extracting from these records any information which is at all commensurate with the opportunities afforded for observation in ancient times. The lion occupied an exceptional position then as now; he was a favorite subject for poetic allusion, for epigram, and rhetorical flourishes. But his character was as much a conventional one at that time as it is at present. This may be also seen in art, where, whether sculptured and painted, or set in mosaics, he was depicted in what were supposed to be characteristic attitudes from Persepolis and the rock tablets of Kaf to the Sea of Darkness, and from the banks of the Orontes to the cities of Africa. He impressed antiquity as he has done the modern world, and so far as disposition and personal qualities are concerned, most of what was known or thought then might have been condensed into the modern statement of his traits given in the French "*Cyclopédie*"; namely, that he was "*si fort et si courageux, qu'on l'a appellé le roi des animaux.*"

What amount of truth there is in this view we shall see; in the mean time it is natural enough to regret that those who might have accomplished so much, have in fact done so little. Varro, Columella, Aulus Gellius, and others wrote on game and hunting, but classic notices of a *venatio* in the amphitheatre are as terse and colorless as entries in a log-book. Marsian boars, or wolves from the Apennines were the most formidable creatures an ancient Italian could find in his own country, and Virgil congratulates himself that such was the case. "*Rabidæ tigres absunt et sæva leonum semina.*" But the scribblers in prose and verse who expatiated upon fish-ponds, nets, gins, snares, Celtic, Lycaonian, and Umbrian hounds, with all the appliances of petty sport, where were they while the *Ludi Circenses* were going on? How was it that these men, who gossiped about everything, never chatted with the keepers of that great *Vivarium* near the Præenestine gate, where there were often wild beasts enough to stock the menageries of the modern world? Why did they not tell of the fleets laden with such cargoes that came to Ostia, interview the men who brought them as they drank rough Massic together in the taverns under the Janiculum, or report the talk of those dark satellites who guarded the *vivaria* of the Colosseum or theatre of Marcellus?

The reason was this: independently of everything else, a Roman of those days was satiated with the sight of actual slaughter until all that now fascinates the attention and enthralls the interest of a reader of adventures had become insipid. The *bestiarii*, or wild beast fighters, were a class apart from other gladiators. So far as our meagre supply of information goes, these men did not meet a royal tiger as a Ghoorka now does; that is to say, did not trust to perfect nerve, training, and activity, to avoid the brute's onset, and slay it by striking at advantage; they appeared in armor and *actually fought* with sword or spear. Considering the style in which lions and tigers combat, one cannot divine the use made of any defensive panoply, which, so far as we can judge, would seem to have been more of an encumbrance than an aid. An iron sword two feet long (for the much-talked-of Iberian steel was most likely only a good quality of untempered metal) could hardly have availed a hampered man in a hand-to-hand struggle of this kind, except in case of accidents that must have been of rare occurrence. Julius Cæsar's Thessalian horsemen chased giraffes around the arena until they were exhausted, and then killed them with a dagger thrust at the junction of the spine and head; but it is safe to say that no *bestiarius* armed with a *venabulum* went through any performances of this kind with a black rhinoceros. Yet every formidable animal on earth perished upon "a Roman holiday." That is, however, all we know.

It is now the fashion to say that lions are such timid

creatures that they might be expected to do little injury if they got out of their cages in the presence of a crowd. When, writes Plutarch, the city of Megara was stormed by Calamus, their keepers or the authorities loosed those lions kept for the games — " opened their dens, and unchained them in the streets to stop the enemy's onslaught. But instead of that they fell upon the citizens and tore them in such a manner that their very foes were struck with horror." Another curious comment upon the timid and retiring behavior of these animals is found in the fact that while they were protected in Africa (preserved for the spectacles) by cruel game laws which deprived people of the natural right of self-defence, the loss of life in that province was so great that it excited compassion even in Rome, and finally led to the mitigation of these statutes by Honorius, and their final abolition during the reign of Justinian.

Moffat ("Missionary Labors and Scenes in South Africa") had the reputation of knowing more about lions than almost any one else, and it was his opinion that eying them was a very questionable proceeding. Both he and Andersson ("The Lion and the Elephant") say that this experiment may sometimes apparently succeed, but "under ordinary circumstances" a hungry lion "does not spend any time gazing on the human eye . . . but takes the easiest and most expeditious means of making a meal of a man." It is not very often that things so arrange themselves as to give any one a chance to try what effect can be produced in this way; still everything that could happen has happened, and combining what follows with the statements already made, it would appear that this much-talked-of personal power is a delusion.

"A lion," writes the Hon. W. H. Drummond (" The Large Game and Natural History of South and South-east Africa ") " will seldom stand much bullying. He may and often will get out of your way, nay, even leave his prey if you approach it, and should you follow him, will perhaps do so a second time, but that is about the extent of it." If interference is pushed further, the lion, "if a male, growls deeply, and makes his mane bristle up round him ; or, if a lioness, crouches down like a cat, lays her ears back, and shows her teeth, and in most such cases, when the brute is fairly roused, a charge is inevitable whether you advance or retreat." On the other hand, " some lions make a point of attacking every human being they meet, without provocation or apparent cause." This is unusual, but " there are many instances of lions having evidently attacked a human being from no other cause than surprise or fear at suddenly finding themselves so close to him. . . . In the above cases, utter immobility and coolness will often avert an attack ; for if the animal, judging by your behavior, imagines that you do not want to hurt it, it will, after trying you for several minutes, and even making one or two sham charges, often walk away and allow you to do the same. . . . Several instances of this have occurred within my own knowledge. A large native hunting party had gone out and were scattered among the thorns, when one of my gun-bearers, who had accompanied it, suddenly found himself face to face with a full-grown male lion, without a yard between them. He had presence of mind sufficient to stand perfectly still, without attempting to take one of the spears he carried in his left hand into the other, and after a couple of minutes the brute walked away, turning its head round every second to watch him.

"This could not be attributed to the efficacy of the human eye, as the man afterwards told me that he had not dared to raise his from the ground. This lion before going far met another native, who raised his spear, as if to throw it; upon which it instantly sprang upon him, and inflicted such wounds that he died within half an hour. I have no doubt that if this man had stood still, he would have been perfectly safe."

A still more striking example of the fact that lions, unless hungry, enraged or alarmed, often pass man by is given by Drummond as follows: "A hunter of mine was following the trail of a herd of buffalo through some dense thickets, alone, and armed only with a single barrel. Suddenly a male lion rose out of one of them, and sitting on his hind quarters, snarled at him; he had hardly seen it when another, about three-quarters grown, showed itself within a few yards on one side, while from behind he could hear the low rumbling growl of a third. Partly turning so as to watch them all, he saw that the latter was a lioness, and that three cubs not much larger than cats were following her. He had, unawares, got into the centre of a lion family. Unfortunately, one of the cubs saw him, and without exhibiting the least fear, ran up to him; upon which its mother, in terror for her offspring, rushed up, and, as he afterwards described it, fairly danced round and round him, springing to within a yard of him, sideways, backwards, and in every way but on him. Luckily he was a man of iron nerve, and bred from the cradle in scenes like this; he therefore remained quiet, taking no more notice of the frantic behavior of the lioness than if she had not existed; for, as he said, it was a hundred to one that I did not kill the mother, and, if I had, the other two would have avenged her." It ended by her ultimately retiring into the thicket, and watching him as he cleared out; but there can be no doubt that any hesitation, nervousness, or involuntary movement on his part would have been fatal.

In his description of the lion, Buffon ("Histoire Naturelle") has delivered a number of opinions based upon imperfect knowledge. This animal, he says, owes its characteristics to climate alone. Lions only inhabit tropical countries, and among the denizens of hot latitudes they are "*le plus fort, le plus fier, le plus terrible de tous.*" On the Atlas Mountains, where snow sometimes falls, these beasts have neither the strength, size, courage, nor ferocity of those who roam the southern plains, and for the same reason, the lion of America, if it deserves that name, is but an inferior beast. Man has greatly circumscribed the range of *Felis leo*, and the natural character of existing varieties has been greatly changed through his inventions. Formerly lions were bolder than they are at present; still, in the Sahara and other places, it happens that "*un seul de ses lions du désert attaque souvent une caravane entière.*" Owing to its brave and magnanimous character, a lion only takes life when compelled to do so by hunger. Certain moral qualities may be said to inhere in the species at large, but there are also individual lions that add to these endowments of their race the finest personal traits. More than one species of this genus exists, and an average lion is about twelve or thirteen feet long. He is less keen of sight, and has not so good an organ of scent as other beasts of prey, and for this reason lions make use of jackals in hunting. All animals they pursue live upon the ground, and in consequence it is not customary with them to climb trees like the tiger and puma — "*il ne grimpe pas sur les arbres comme le tigre ou le puma.*" Their attack is always made from an ambush, whence the victim is sprung upon and struck down; but it is not devoured until after life is extinct.

All this, it may be repeated, is erroneous. Climate alone does not form geographical varieties. Species require to be adjusted to the whole physiography of their respective regions, and to their organic environments as well. The lion inhabits temperate latitudes where the weather is often cold, and it is on those parallels which in Africa run north and south of the equatorial belt, that he attains his highest development.

With respect to the lion of the Atlas, Major Leveson ("Hunting Grounds of the Old World"), General Daumas ("Les Chevaux du Sahara"), and Gérard ("Journal des Chasseurs") have shown that it is larger than its congener further south. Buffon's thirteen feet lions belong to an earlier geological period than ours; no such specimens of the cat kind are at present alive, but his tribute to the courage of the king of beasts is not perhaps altogether undeserved. Of course there is nothing in his remarks about magnanimity and the like, and as for a single lion attacking a caravan, the statement is absurd. Lions and troops of lions are described by many observers — Le Vaillant, Cumming, Oswell, Harris, Davidson, Kerr — as having forayed upon encampments in various ways, but there is no authentic account of any incident such as Buffon relates.

What he says about the animal's deficiency in sight and scenting power is not supported in any way by facts. There is nothing in the creature's anatomy to warrant such an assertion. Its olfactory apparatus is well developed, and as it is a beast of prey, and belongs to a family distinguished for keenness of scent, there is no reason to think that this function does not correspond with its structure. Neither is there anything, so far as the writer knows, in the better class of observations made upon lions, to indicate any deficiency in this respect. With reference to sight, if Buffon meant more than that they, as being nocturnal in habit, are at a disadvantage in the sun's glare, it was, we must believe, a mistake upon his part. Their organ of sight is structurally of a high order; it is so placed that the range of vision is large, and no good authority has disparaged the lion's far-sightedness, or the defining power of his eye.

None of the great cats is, however, strictly nocturnal except in places where they are constantly pursued. Lions frequently stalk or drive game while the sun is up; they see perfectly well during these hours, and it is evidently a mistake to give the primary importance commonly attributed to it to a peculiarity of vision which the *Felidæ* have in common with other classes.

Buffon's opinion of the use to which lions put jackals falls to the ground before facts. It is an old idea that they, and tigers also, employ them as scouts; nevertheless it would appear that the true relation has been overlooked, and that it is the jackal who uses the lion. When a lion leaves his lair he always roars, and if any jackals are in the vicinity, the sound attracts them at once; it is like an invitation to a meal, for these satellites feast upon the offal. Similarly, as the lion's majestic form moves with long and soft but heavy tread through the gloom, every jackal that sights the grim hunter follows him.

In works on natural history lions are classed among the *educabilia*. There is, however, a certain ludicrousness in distinguishing this animal as one that can be taught. So can a flea. Every creature with a nervous system may be and is instructed in some manner. All living things so provided learn, though not necessarily through tuition, nor in all cases consciously. Dr. Maudsley's remark ("Physiology and Pathology of the Mind"), that "a spinal cord without memory would be an idiotic spinal cord," is full of meaning. Wherever a nervous arc exists, there is memory and the potentialities of mind. The central axis is nothing more than an integrated series of such connected arcs ending in a brain when the animal is sufficiently elevated.

A whelp is born in the spring, or towards the close of winter, a little sooner or later, as the latitude varies. Before this event the parents have fixed upon some solitary spot in which to establish themselves. The mother's character undergoes a temporary change for the better during the period of maternity. While the pairing season lasts she is a shameless wanton, ready at any moment to abandon her mate for a stronger rival. Desperate combats accompany the lion's courtship, in which both parties are frequently killed, and in almost all instances these are brought on by the lioness, who seems to take a savage pleasure in provoking such duels.

Gérard gives the following story, which is in all essentials a true picture of the behavior of both males and females at the time spoken of. "It was in the stags' rutting season, and Mohammed, a great hunter of every kind of wild animal, perched himself at sunset in the boughs of an oak tree to watch for a doe that had been seen wandering in the vicinity, accompanied by several stags. The tree he climbed stood in the middle of a large clearing, and near it was a path which led into the neighboring forest. Towards midnight he saw a lioness enter this open space, followed by a red lion, with a full-grown mane, who carried the carcass of an ox. Soon after they were followed by another lion, a lioness, and three cubs. The first lioness strolled from the path, and came and laid herself down at the foot of the oak, while the lion remained in the path, and seemed to be listening to some sound as yet inaudible to the hunter.

"Mohammed soon heard a distant roaring in the forest, and the lioness immediately answered it. Then the lion commenced to roar with a voice so loud that the frightened man let his gun fall, and held fast to the branch with both hands lest he should tumble from the tree.

"As the voice of the animal heard in the distance

gradually approached, the lioness welcomed him with renewed roarings, and the lion, restless, went and came from the path to her, as if he wished her to keep silence, and then, from the lioness to the path again, as if to say, 'Let the vagabond come; he will meet his match.'

"In about an hour, a large lion as black as a wild boar stepped out of the forest and stood on the edge of the clearing in the full moonlight. The lioness raised herself up to go to him, but the lion anticipating her intention, rushed before her, and marched straight towards his adversary. With measured steps and slow they approached to within a dozen paces of each other; their great heads high in air, their tails slowly sweeping down the grass that grew around them. They crouched to the earth; a moment's pause, and then they bounded with a roar high in air, and rolled upon the ground, locked in their last embrace.

"Their struggle was long and fearful to the involuntary witness of this midnight duel. The bones of the combatants cracked under their powerful jaws, their talons strewed the grass with entrails, and painted it red with blood, and their roarings, now guttural, now sharp and loud, told of their rage and agony.

"At the beginning of the conflict the lioness crouched low, with her eyes fixed on the gladiators, and all the while the battle raged, manifested, by the slow, cat-like motion of her tail, the pleasure she felt at the spectacle. When the scene closed, and all was still and quiet in the moonlit glade, she cautiously approached the spot, and snuffling at the bodies of her two lovers, walked leisurely away, without deigning to notice the gross but appropriate epithet Mohammed sent after her, instead of a bullet, as she went off."

This otherwise excellent sketch loses something of its *vraisemblance* from carelessness and inaccuracy in execution, and also from an unfortunate style, which gives to most French narratives of this kind, however true, the air of romances. Gérard knew that a doe is never accompanied for any length of time by several stags, and there can be no excuse for making a lion range the woods with an ox in his mouth.

When cubs are about two months old, they begin to forage in the vicinity of their lair. This hunting, however, is more than half play, for they are sprightly little creatures whose gambols and infantile familiarities soon become distasteful to the grave and morose nature of their father. The lion then takes up his quarters out of their reach, but at the same time near enough to come to the assistance of his family if aid should be needed. Two cubs as a rule are born together, and one of these is generally a male. If the birth be single, this is said to be invariably the case, so that the fact that males considerably outnumber females is accounted for, and with it both the wantonness of the latter, and those trials to which their consorts are exposed. The race maintains its place by the sacrifice of its weaker numbers. The strongest whelps and most powerful lions live, mate, and kill or dispossess their rivals. Sexual selection on the lionesses' part aids this process, and the result is, as everywhere and always, that the fittest survive, and transmit their traits with a result which is in every way beneficial to the species.

A great many young ones die while cutting their teeth. If this has been accomplished safely, however, their education begins immediately after that event.

A lion does not reach maturity until the eighth year, and he lives to be about forty. At the end of his second year, however, the animal has attained considerable size, strength, and agility, while his predatory tendencies are then more freely indulged than at any subsequent period of life. Up to the time at which mutual indifference separates parents and offspring, the latter have been directed and assisted in all things. Game has been found for them, and methods of capture and killing have been illustrated. Thus far experience has brought with it only assurances of success. They have been incited to take life for practice, encouraged to act when there was no necessity for acting, guarded from the consequences of temerity and incapacity. Therefore, when separation takes place and they go forth alone, it is with an undue self-confidence which often entails disaster. Young lions are notoriously daring, destructive, and dangerous.

There are many dogmatic and differing decisions with regard to the manner in which lions seize, kill, and eat their victims, as also in respect to the degree in which their natural ferocity may be tempered by fear or discretion. There must be, of course, a family likeness among them in these particulars, but no such uniformity as has been imagined can be found in their behavior when a wide enough view is taken.

The fanciful opinion that a lion disdains to eat game that he has not stricken himself, vanishes at once. Derogatory to his dignity as it may be, the fact is that he will consume anything he finds dead, that his taste is of the most indiscriminate character, and that he is very frequently a foul feeder. "Many instances," says Andersson ("The Lion and the Elephant"), "have come to my knowledge which show that when half famished he will not only greedily devour the leavings of other beasts, but even condescend to carrion." In another work ("Lake N'gami") the same author states that lions eat carrion without being "half famished." Sir Samuel Baker ("Nile Tributaries of Abyssinia") saw several that he knew were not pressed by hunger feeding on the putrid body of a buffalo shot by himself, and Gérard ("Journal des Chasseurs") very nearly lost his life by a lioness who had come to feed upon the carcass of a horse in the last stages of decomposition Lions appropriate any meat they may happen to find. "I have frequently discovered them feasting on quadrupeds that had fallen before my rifle," remarks Colonel Cumming ("A Hunter's Life in Africa"). Major Leveson ("Sport in Many Lands"), W. H. Drummond ("The Large Game and Natural History of Southern Africa"), Colonel Delgorgue, ("Voyage dans l'Afrique Australe"), Sir W. C. Harris, ("Wild Sports in Southern Africa"), and H. C. Selous ("A Hunter's Wanderings in Africa"), all confirm the assertion that "lions are by no means too proud to eat game killed by others." This charge must be admitted, and it is entirely conformable with another; namely, that his majesty is one of the laziest beings alive. "Laziness, assurance, and boldness," says Gérard, are his most conspicuous traits of character, and Moffat ("Missionary Labors and Scenes in South Africa") adds gluttony to the list. He was "taken aback," he assures us, by the astonishing feats in the way of gormandizing that this animal performed. It should be remembered, however, than an average beast of prey passes a life divided into alternate periods of famine and repletion, and that it is, both from habit and conformation, capable of cramming itself in a manner which almost exceeds belief.

There is hardly need to cite authorities upon the act of seizing prey, because lions do so in all those ways that different observers have severally decided to be peculiar to this beast; and it is the same with the various methods by which they kill. The whole subject of attack, whether upon man or beast, is wrapped in a mass of positive contradictions.

In India troops of lions have been known to divide themselves into sections that relieved one another at short intervals in the actual pursuit of game. As a rule, however, species belonging to this group do not, and can not, really run down prey. Their peculiar structure, adapted to bounding, climbing, and brief rushes, does not admit of a long gallop. Their limbs are too massive and short, and are not sufficiently detached from the body to give them free play. Lions have been called "the most cat-like of all cats," and for the most part these animals ambush or stalk those creatures which they kill.

When a lion impelled by hunger leaves his lair, he sometimes has a definite object in view, but more frequently goes forth to take advantage of anything that may turn up. If the former is the case, his course is directed, as that of a man would be in like circumstances, by a previous acquaintance with the haunts and habits of the game he is after. He does not ambush a disused path to a dried-up spring, or look for a quagga in a buffalo wallow, or attempt to stalk black antelopes in the same way that he would approach cattle belonging to some Hottentot kraal.

In Africa, which is his true home, a lion "is never known to chase prey." Having sighted it, ascertained its species, surveyed the ground, found out the direction of the wind, — preliminaries essential to any subsequent attempts to get near, — he begins to practise a set of manœuvres adapted to present conditions, and these he has learned in the literal meaning of that term. Faculty is transmitted. Knowledge is always acquired.

Having closed successfully and seized his prey, it is destroyed in a variety of ways. As a matter of fact immediate death does not invariably come to the relief of its sufferings, even in the case of those smaller creatures on which the lion preys. He does not wait, as Buffon supposed, until insensibility ensues before tearing them to pieces. Nor is it true, as Dr. Livingstone imagined, that Providence assuages the agonies of all animals thus caught, by bestowing upon the *Felidæ* a propensity to shake their victims, and so produce a state of insensibility. How can a lion shake an ox or an eland, a horse, giraffe, buffalo, or young rhinoceros? Andersson tells us that he mistook the groans of a zebra carried past his camp by night for those of a human being, and went to the rescue. More than this, if the brute itself has any feeling about this matter,—and there is every reason to believe that it has,—all manifestations of pain heighten the pleasurable excitement it experiences in putting an animal to death. Cruelty is organized in its brain, and to a beast of prey, pity is about as possible as poetic inspiration. Love of bloodshed, exultation in carnage, immitigable ferocity, are ingrained in them all; and so far as a lion appreciates expressions of mental anguish and physical torture, they thrill his fierce spirit with a savage joy.

Gordon Cumming relates a story which shows what a human being may experience when in the clutches of a lion. His party had encamped, and "the Hottentots," as he tells, "made their fire about fifty yards away, they, according to their custom, being satisfied with the shelter of a large bush. The evening passed away cheerfully. Soon after dark we heard elephants breaking trees in the forest across the river, and once or twice I strode away into the darkness, some distance from the fireside, to stand and listen to them. I little, at that time, dreamed of the imminent peril to which I was exposing my life, nor thought that a blood-thirsty, man-eating lion was crouching near, and watching his opportunity to spring into the kraal and consign one of us to a horrible death. About three hours after the sun went down I called to my men to come and take their coffee and supper, which was ready for them at my fire. After supper three of them returned before their comrades to their own fireside, and lay down; these were John Stofolus, Hendric, and Ruyter. In a few moments an ox came out by the gate of the kraal and walked round the back of it. Hendric and Ruyter lay on one side of the fire, under a blanket, and Stofolus lay on the other. At this moment I was sitting, taking some barley broth; our fire was very small, and the night was pitch dark and windy.

"Suddenly the appalling and murderous voice of an angry and blood-thirsty lion burst upon my ear within a few yards of us, followed by the shrieking of the Hottentots. Again and again the murderous roar of attack was repeated. We heard John and Ruyter scream, 'The lion! the lion!' Still, for a few moments, we thought he was but chasing one of the dogs round the kraal. But the next instant Stofolus rushed into the midst of us almost speechless with fear and horror, his eyes bursting from their sockets, and shrieked out, 'The lion! the lion! he has got Hendric; he dragged him away from the fire beside me. I struck him with burning brands upon the head, but he would not let go his hold. Hendric is dead! O God! Hendric is dead! Let us take fire and seek him!' The rest of my people rushed about, yelling as if they were mad. I was angry with them for their folly, and told them if they did not stand still and be quiet, the lions would have another of us, for very likely there was a troop of them. Then I ordered the dogs, which were nearly all tied, to be loosed, and the fire increased as far as it could be. I shouted Hendric's name, but all was still. I told my men that Hendric was dead, and that a regiment of soldiers could not help him then. Hunting my dogs forward, I had everything brought within the kraal, when we lighted our fire, and closed the entrance as well as we could.

"My terrified people sat around the fire with guns in their hands, fancying at every moment that the lion would return and spring into the midst of us. When the dogs were first let go, the stupid brutes, as dogs often prove to be when most needed, instead of going at the lion, rushed fiercely at one another and fought desperately for some minutes. After this they got his wind, and going at him, disclosed his position. They kept up a continual barking until day dawned, the lion occasionally springing at them and driving them in upon the kraal. This horrible monster lay all night within forty yards of us, consuming the wretched man he had chosen for his prey. He had dragged him into a little hollow at the back of the thick bush beside which the fire was kindled, and there he remained until day broke, careless of our proximity.

"It appeared that when the unfortunate Hendric rose to drive in the ox, the lion watched him to his fireside, and he had scarcely lain down before the brute sprang upon him and Ruyter (for both were under one blanket) with his appalling roar; and, roaring as he lay, grappled him with his fearful claws, and kept biting the poor man's chest and shoulder, all the while feeling for his neck, having got hold of which, he dragged him away backward round the brush into the dense shade.

"As the lion lay upon him he faintly cried, 'Help me! Help me! Oh God! men, help me!'"

Here was no instinctive fear of man, no sign of the timidity so much talked about, no falling off of the victim into the dreamy languor Dr. Livingstone expatiates upon. His pain was sooner over than that of some we know of; death came when the neck was crushed, but what had he suffered previously?

There is an alleged trait of character which should be alluded to on account of the propensity displayed even by those who really know this animal to make a composite being of him — part lion and part gentleman.

Gérard is one of them. He was to some extent, no doubt, deceived by common report, and likewise misled by his knowledge of those domestic virtues that really belong to the animal. At all events he constructed a lion that bears a curious resemblance to a *raffiné* of the famous old duelling days in France without the *seigneur's* levity or his lewdness. When his family, whom he has up to this time fed himself, are able to join in the chase, the lion finds the game, strikes it down, and then, with that refined self-abnegation which comports so well with his natural character, he retires to a little distance from the quarry in order that *Madame* may be first served. This and much more to the same effect.

It happens, however, that one man, and only one to the writer's knowledge, the Hon. W. H. Drummond, chanced to see what Gérard has depicted in colors furnished by his own fancy. His narrative of the incident from first to last is much more in accordance with the style of manners taught in the struggle for existence than the former one. One day while watching the motions of some antelopes from the summit of a grassy and rock-strewn ridge, Drummond suddenly became aware that he was not the only hunter interested in the game of that vicinity. A lioness with her whelps crouched among the herbage at a little distance, and so intent were they upon the movements of their expected prey, that he was entirely unnoticed. While awaiting events a band of quaggas passed close to some bushes at the foot of the slope, and then a lion's form was launched upon the leading stallion, and he fell dead from a blow with the beast's forearm.

Without any delay the lion proceeded to help himself, his family drawing near, but waiting until his appetite had been stayed. "The sultan of the desert" has a short temper when he is feeding, and on many occasions has been known to eat his wife, either in the way of reproof to her importunity at such times, or because he did not have food enough. It would seem that this lioness suspected something of the kind might occur, for she kept herself and the young ones in the background until his highness had finished, which he, not being particularly hungry, did very soon. When he had walked away and stretched himself out, the rest pressed forward, and the mother treated her offspring with scant curtesy. She pounced upon those parts she preferred, and boxed the little ones, who were struggling for a bite, out of the way whenever they incommoded her.

Thus far in the catalogue of leonine gifts and graces we have not discovered any that are peculiarly their own; on the contrary, when examined closely, those with which lions are accredited turn out to be counterfeits. Gordon Cumming says of the lion, in company with his mate and whelps, that, "at this time he knows no fear," and in defence of his family "he will face a thousand men." This is a rhetorical flourish, and yet now when it has become the fashion to call the creature a poltroon, the statement as it stands is better supported by proof than almost any other that has been made concerning its character. If this animal is not brave, nobody is in a position to call it cowardly. All the evidence tends the other way. Taken as it is, looked upon as a brute to whom heroism, sentiment, and high resolve must be as impossible as righteousness, the lion preserves the demeanor of courage better than any other member of the *Felidæ*.

Moffat, Lichtenstein, Freeman, Rath, Galton, say with W. C. Kerr ("The Far Interior") that "when a lion is thoroughly hungry there is no limit to his audacity and daring." Every being must have *some* incitement to action, and those motives which are most powerful with lions appear to be anger and appetite.

Postponing for the moment his relations with mankind, let us see what kinds of game the lion is accustomed to prey upon. No coercion can be exercised in this direction. Actual starvation might take away liberty of choice, but, as a rule, it must be admitted that a selection of this kind is significant of the opinion which an animal has of its own powers, as it also is of its boldness. The giraffe, which lions occasionally kill, is entirely defenceless : so with elands and all antelopes. This is likewise the case with those domestic animals which are devoured. It has been said that the elephant is sometimes attacked, but this is one of those stories which only display the ignorance of those who propagate them. The black and white rhinoceros is never assailed, although Delgorgue actually refers to the latter, a beast second only to the elephant in size and most formidably armed, as if it were commonly destroyed. "*Maintes fois trouvai-je des rhinocéros de la plus haute taille, que ni leur poids, ni leur force, ni leur fureur, n'avaient preserver de la mort.*" If anything were needed to set off this pleasant statement, it could be found in Delgorgue's roundly declared opinion that lions are all "abject cowards."

But in Africa the lion constantly preys upon the buffalo, and without going so far as Andersson in saying that he principally lives on this species, the fact that it is continually killed is beyond question. Many famous hunters suppose that an African buffalo is the most dangerous creature to be found on the "Dark Continent." It is of immense size and strength, active, brave, and fierce.

No account is known to the writer of a single lion that was seen to slay a full-grown buffalo, and several authorities doubt whether this be possible. The latter have, however, often been shot while bearing the scars of combats with one or more lions. According to the evidence as it exists, the case stands in this way. One lion may attack a buffalo, it is impossible to say whether he will or not; two of them certainly do so, and the battles that ensue are of the most desperate description. It is known, also, that these conflicts do not always end in favor of the assailants.

"The lion kills only for food," says Major Leveson, meaning that in mature life he does not commit useless murders, or show the same love of blood for its own sake as some other members of his family. Without doubt, this animal is not sanguinary when compared with a panther or puma, but it is quite as likely that he is restrained from unnecessary carnage by economic views, as by any sentiment of generosity or mercy.

A lion when surprised does not usually dash away incontinently; if his retreat is not interfered with, and he has learned that firearms are more effective than his own weapons of offence, he falls back slowly. When so placed that they cannot escape, some lions die like curs, but the majority of accounts represent them as perishing gallantly. Such is the case also when for any reason the creature has resolved to fight. Then it seems to make no difference to him how many foes he encounters. Numerous narratives very similar in detail have been written by different observers of such scenes. No other wild beast confronts a body of armed men after his manner. That last parade in face of a horde of savages beneath whose assagais he is about to die, is so striking that false inferences from the sight can scarcely be avoided. It is not the "deliberate valor" of Milton we see, nor even heroic despair; it is nothing perhaps with which humanity in its nobler emotions can sympathize; but it looks as if it were, and men have yielded to their feelings and believed that it was. "Life," says Professor Robinson, "has but one end for a lion — enjoyment. He is incapable of forgetting that he is only a huge cat, or flying in the face of nature by pretending to be anything else. . . . He makes no claim to invincible courage; on the contrary, he prefers, as a rule, to enjoy life rather than to die heroically. But when death is inevitable, he is always heroic, or even when danger presses him too closely . . . a lion in the shadow of death remains a lion still."

All things being equal, lions conduct themselves towards mankind according to the suggestions of the time being and their previous experiences. One that had just eaten an antelope might pass by a man; another might kill him. The former, by all accounts, is the more likely to occur, and it is said that Bushmen and other natives can tell by the voice whether he is full or fasting; and in the first case have no fear that he will become aggressive without provocation. When forbearance is not a matter of repletion, it is no doubt, in some measure, the result of sloth. A lion never does anything he can avoid doing.

Baker's story of the lion that met a Nubian sheik with two companions, and tore the leader to pieces, is one of a great number of instances that might be brought forward to show that wherever these animals are not conscious of being put entirely at a disadvantage by superiority of arms, they display little of that fear of man which is commonly attributed to them. Poorly-armed tribes are under no such delusion. The Ouled, Meloul, or Ouled Cassi Arabs whose *douars* were attacked would have been as difficult to persuade of the lion's timidity towards mankind, as those Makubas on the Ghobe, or "the miserable Bakorus," whom he devoured at his good pleasure. Dr. Schweinfurth ("The Heart of Africa") was at an Egyptian garrison where the soldiers were carried off from within their own lines night after night. Moffat, Delgorgue, Livingstone, Cumming, all record incidents of what they call his "desperate attacks." Still, and as if to show what it is possible for men to commit themselves to when writing about wild beasts, we have Burchell's opinion (" Travels in the Interior of Southern Africa ").

This author, according to his own account, spent four years in a lion country, and saw but one during the whole of this time. That one was accidentally encountered on a journey, and they succeeded in shooting it through the body, upon which it drew off into the bushes and disappeared. Yet it is on the strength of an experience like this that Burchell says he has "no very high opinion of the lion's courage." Of course the reference has an appearance of being overstated, but whoever reads the bulky quartos in which these travels are written will find that such is not the case.

So much in the way of a review of Buffon's general description.

It is easier, however, and safer to decide as to what lions are not, than to say what they are. Almost everything written upon this subject deals nearly to exclusion with the animal's habits, and leaves its character untouched. Even in this respect also our information is not complete.

C. J. Andersson ("The Lion and the Elephant") remarks that "the modes of life" belonging to "the Lord of the African Wilds" are not at all thoroughly known, and he expresses an opinion fully justified by facts to the effect that he has himself been able to bring together much information in this connection that "may not have been noticed by other travellers and sportsmen."

In making up a summary of what has gone before, the writer is much indebted to this valuable work.

We have no psychological scheme for lions, and must take their characteristics as they happen to present themselves, without any pretence at arrangement, based either upon their natural order or real importance. There is an account given in MS. to Lloyd, the editor of Andersson's posthumous papers, that shows the character of the Indian lion in much the same light that his African congener has been placed by Baker, Drummond, etc.

"This beast was believed to have his lair in a patch of copse-wood where, from the jungle having been some years previously cut away by the natives for stakes and the like, the young trees had grown up again so close and tangled as to be almost impenetrable. But this patch was of no great extent, its area, perhaps, not exceeding that of Grosvenor Square. The other parts of the wood surrounding the tank were in a state of nature, consisting of bushes and timber trees.

"On reaching the ground, the natives were stationed in the trees thereabouts as markers. But it was not till the party had beaten the patch with their elephants for a considerable time that the lion was discovered to be on foot, and some further time elapsed before he was viewed as he was stealing away from the brake, along a sort of hedge-row, for the more open country beyond. Captain Delamaine, who was some forty or fifty paces from the beast, then fired, and wounded him severely in the body.

"On receiving the ball, the lion immediately faced about, and charged my elephant, but the nerves of the latter having been recently shaken by wounds received from a royal tiger, turned tail, and regularly bolted. In the scurry through the jungle, one of the guns, having been caught by a tree, fell from the howdah and was broken, a loss, as the sequel proved, that might have been attended with very disastrous consequences."

But the lion soon gave up the chase, and retraced his steps to the patch whence he had been started. Here he was followed by Captain Harris alone, Delamaine's elephant, from its late fright, having become too unsteady to be taken into thick cover.

"The Captain soon found and fired at the beast, which in its turn instantly sprang at, and made a fair lodgment on the head of his elephant, but the latter being a large and powerful animal, and accustomed to the *chasse*, almost immediately shook off its fierce assailant, who fell with violence on the ground." This desperate mode of attack and reprisal was on both sides repeated in more than one instance, and this, moreover, within view of his companion, who, though prevented — for the reason mentioned — from taking part in the conflict, was, from the outside of the brake, intently watching the proceedings of his friend. After a time, whether because he left the patch, or from having concealed himself, the beast was no longer to be found.

"It was at the period of the monsoon, and just as the hunters were at fault, there came on a heavy shower of rain, when, principally for the sake of the guns, it was deemed best to retire for shelter to some trees in the more open country at a few hundred paces distance.

"The storm soon passed over, but being doubtful whether their guns might not be wet, it was thought advisable to discharge them. This was no sooner done, however, than the lion began to roar terribly, and continued doing so for some time, in the direction of the late scene of conflict, from which it was pretty evident, that, though they had been unable to find him in the patch, he had been harbored there the whole time.

"When reloaded, the party therefore returned to the brake, and were informed by one of the markers that on the report of the guns, the lion had rushed roaring from it into the more open country, evidently for the purpose of venting his rage on the first object that came across his path. On proceeding a little further they were hailed by another marker, who told them that the brute was crouched in a cluster of brambles, of a very limited extent, about twenty paces from the very tree in which he himself was perched.

"As the country was pretty open around the thicket in question, the sportsmen were able to reconnoitre it narrowly, and that without taking the elephants into the very thick of it, which was deemed unadvisable, as, had those animals come directly upon the lion, they might have been scared and rendered unmanageable. But the beast was not perceptible.

"From the cover being so limited in extent, it appeared to be almost an impossibility that the lion could be there, the rather that the elephants, so remarkable for their fine sense of smell, did not seem at all aware of his presence, and it was in consequence imagined that the man must be mistaken. But as he persisted in his story, it was determined to fire a shot into the thicket, which was accordingly done, though without any result.

"When a lion, that has been wounded and hotly pursued, has 'lain up,' or hidden himself, for a time, his position is generally known by his roaring, panting, or hard breathing; but in this instance there were no indications of the kind, which, coupled with the shot having failed of effect, confirmed their previous impression, and they were, therefore, on the point of moving off elsewhere.

"But as the marker continued asseverating from his tree that the brute was positively lying in the very brake near which they were standing, it was resolved to try another shot, which was fired by Captain Harris' man, who was seated at the back of his master's howdah.

"This had the desired effect, for the gun was hardly discharged, when the lion, with a tremendous roar, sprang up from his lurking-place, and in a second was once more on the head of Captain Harris' elephant. But he was almost immediately shaken off, when he retreated to the same brake from which he had issued, and where, as before, he was no longer discernible.

"A shot was therefore directed towards the spot where he was supposed to be, and he again charged the Captain's elephant, and on being dislodged trotted off towards the patch that harbored him in the first instance.

"During the *mêlée* just described, Captain Delamaine, from an apprehension of hitting some one, had been deterred from firing; but as the lion was retreating he discharged both barrels of his double gun, and broke one of the hind legs of the beast.

"Upon receiving this wound the lion at once turned, and rushing at the elephant, sprang up on his hind quarters and fixed his fangs in the thick part of the tail. The poor animal perfectly screamed from the extreme torture, which was little to be wondered at, as this unfortunate appendage had only a week previously been severely lacerated by a huge tiger. The elephant now swayed to and fro to such a degree that his rider had some difficulty in retaining his seat in the howdah, and was much less able to take an accurate aim at the lion, which, screened as it was by the protruding rump of the elephant, would have been scarcely practicable. The Captain, besides, had only one barrel remaining, and it therefore behooved him to be most cautious that his last charge was not ineffectually expended.

"This trying scene continued for two or three minutes, during which Delamaine anxiously looked out for Captain Harris. But unluckily his elephant had been rendered unmanageable by the maltreatment it had itself received from the lion, and it was not, therefore, in his power to render aid to his friend."

The appearance of the lion at this time, maddened as he was with pain and rage, is described as most awful.

"At length the beast's long-continued attack on the elephant caused the poor animal evidently to give way and to sink behind, and had the affair continued a short time longer, there is no doubt it would have been on its haunches, and the rider at the mercy of the fierce assailant.

"Finding matters in this very critical state, it became necessary for him to risk everything. Leaning, therefore, over the back of the howdah, and clinging to it with one hand, he with the other discharged his rifle, a very heavy one, at the head of the lion (the piece at the time oscillating, or swinging, in a manner corresponding with the roll of the elephant), and as luck would have it, the ball, after crashing through the beast's jawbone, subsequently traversed the whole length of its body.

"This caused the lion to let go his hold, and for a few seconds he appeared to be partially paralyzed, but recovering himself, he slowly retreated towards the thicker cover."

Subsequently he was again attacked by the party, and in two or three instances charged them as gallantly as ever; but as he was always received with a heavy fire, an end was at length put to his existence.

There is no need to add much to what has been said of the effect produced by inherited and personal experience. Nobody denies that lions are possessed of intelligence, and this being the case, they learn to avoid known dangers, and to take advantage of those conditions which have previously proved favorable. If this and what it implies were not true, there could be but one reason for it, which is that the race was congenitally idiotic. Therefore to dispute about the lion's courage as if there might be archetypal beasts differently endowed from those representatives of their species which naturally, and of necessity, vary in boldness with changing environments, appears to be a waste of time. Furthermore, the possession of power of any kind to a great degree determines its exercise, and it is impossible to suppose that an animal which, above all others, except the tiger, is specialized for violence, will not be blood-thirsty and aggressive.

Sir Samuel Baker appears to be the only writer, really an authority, who knows nothing authentic and has no personal cognizance of the forays of lions upon villages and camps. Delgorgue, Harris, Cumming, Andersson, and everybody else whose opportunities for observation have been at all extensive, recognize such incidents as perfectly well established. Indeed, taking the character of this beast and its situation into consideration, the only thing surprising about the matter would be that it had not done those things upon whose reality Baker seems to cast a doubt. Drummond relates a story in this connection, in the scenes of which he was himself an actor, and as many of those traits which have been discussed are well brought out in his narrative, it is given in full.

"In two cases I have been an accessory to the death of well-known man-eaters, one of which had almost depopulated a district. . . . The locality in which this one committed his depredations was in the northeast corner of Zululand, where a number of refugee Amaswazi had been located, and when I arrived they had continued for nearly a year, so that many villages were deserted, and all had more or less suffered; for the brute did not confine himself to any one in particular, nor come at any regular intervals, but so timed his visits that no one was sure of his or her life from day to day. No fastenings were of any use against him, as his immense strength enabled him to force an entrance if he could not find one ready made, while the outer ring-fence, of interwoven thorns, supported by strong posts, which guards all native villages, and is often of great height, offered no obstacle to his powers of jumping, a single bound being always sufficient to land him inside.

"He usually confined himself to killing a single individual, and would claw one out from under the blanket or skin under which, with covered heads, they cowered in terror on his arrival; but on the two or three occasions in which he had met with opposition, and when he had been wounded with assagais, he had killed every soul in the hut, and so dreadfully mangled them that their bodies almost defied recognition.

"I was staying at the villages for some weeks, first at one and then at another, as they suited the position of the game, or where I happened to find myself at night; but though I heard of the lion having attacked one either just before or just after I had been there, I never happened to meet it, and the ignorant natives became anxious for my presence, saying that their enemy feared to go where I was.

"This, however, was not destined to last. One sultry evening I arrived at the outermost village, having been forced to leave the spoor of a herd of elephants for want of ammunition, and being very tired, I determined to sleep at it, sending on two of my men to fetch some from the place which I made my headquarters. Tired as I was with my exertions on an unusually hot day, I soon fell asleep in the hut that had been given up to our use; but, as the heat was stifling, I was not at all surprised at being awakened towards midnight by a heavy thunderstorm, which crashed round us for half an hour or more. At last the hush came that always accompanies the tremendous rain which follows, and seems to quench such storms, broken only by the heavy splashing of big drops, and the gurgle of the water that flooded the ground, and I should soon have been asleep again had not a drop come splash into my face through the ill-thatched roof, almost immediately followed by a small stream, of which it had been the advanced guard. This necessitated my looking out for a drier spot, when suddenly out of the quiet of the descending rain, came such a confused clamor of shrieks and cries, of yelling and moaning, that until I heard the voice of the lion, I was utterly unable to account for it. This lasted for full half a minute, and then such a blood-curdling scream of mingled pain and despair came as I hope I may never hear again, and which haunted my dreams for many a month after.

"My men, and among them two old hunters, each of whom had killed several lions, shrunk crouching back to the further end of the hut, returning no answer to my words when I told them to come out with me and face the beast, though, as I opened the hut entrance, and looked out on the pitch darkness, it was evident how useless any such attempt would be. The death-yell we had heard was followed by silence for some time, during which the brute was probably departing with its victim, and the natives were still afraid of its return; then the usual noisy lamentations for the dead broke forth, and were continued without intermission until daylight, though I was so tired that, without expecting it, I fell asleep again, and did not wake any more that night.

"There was little to tell when morning did break. The lion had hit upon the most crowded hut of all, the one in which the people who had given place to us were sleeping in addition to its regular owners, and had picked out a young married woman, taking her from among several, without injuring any one else; as they said — 'a man does not stab more than one of his herd of cattle when he is hungry.'

"Previous to this, on my first arrival, the head man of the district had come and asked me whether I would assist him to destroy this brute, as, if so, he would turn out with all his people, and beat up the country until it was found; and in point of fact we had already done this, on the occasion of the chief's uncle having been carried off; but the ground was so dry and hard then that our best spoorers failed to hit off the track. To-day, however, as the rain had ceased a few minutes after its departure, there could be no doubt about finding it, and as soon as I awoke I sent off to the chief to ask him to come with his men, saying that, whether he had arrived or not, I should take up the trail at nine o'clock.

"I did not at this time know that the woman who was the last victim was his relation, but when my messenger came back and told me so, adding that the chief was fearfully angry, it did not surprise me to hear that runners had been sent out already, and that he had threatened to drive out of the country any one old enough to carry a spear who remained behind, and that if I could wait until the sun had reached a certain part of the heavens (till about ten o'clock), he would join me.

"I had already had breakfast when this news came, and to save time I took a hunter and a spoorer (tracker) with me and followed the lion. About two hundred yards off we found the spot where he had made his disgusting meal, and then the track led right away towards a stream, nearly a mile distant, where he had quenched his thirst. Keeping steadily on, he passed through several covers quite strong enough to have held him, and through which we had to pass with the utmost caution, until, at length, he came out on to the open, and headed in a direction that we knew could lead nowhere but to the Umbeka bush, the thickest jungle for miles around. As this was still nearly four miles off, I sent one man back to tell the people where to come to, and kept on with the hunter.

"On reaching the jungle, which covered the entire side of a hill, and was stony and broken to the last degree, besides having its undergrowth formed of impenetrable cactus, we did not of course attempt to enter, but separating, walked round it, the upper and more rugged portion falling to my share, and carefully examined every inch of the ground to see whether by any chance he had again left it ; however, no vestige of his spoor could be seen, and by the time we got back to our starting-point, the whole of Tekwane's people were in sight.

"The chief himself was with them, though he had no intention of taking any active part in the proceedings, and when we started he retired with some of his old men to a place of safety, and a council of how to proceed was held on the spot. My idea was that the guns should guard the more likely passes, while the people, numbering near five hundred, should beat out the jungle. To this, however, the objection was offered, that from the well-known thickness of the place, and the universal terror of the lion, the men would not attempt to beat it unless they were led by myself and my hunters. Such being the case, it was decided that spies should be placed in the tree-tops and other commanding positions, while the great body of the people were to enter at the top and drive down; but knowing as I did how very dangerous the affair would become if the lion was wounded in such cover, in many parts of which one could not see a yard off, I specially ordered my men not to fire unless they felt sure of killing or disabling the brute on the spot, and advised that every one, advancing in as unbroken a line as possible, and going slowly and making all the noise possible, should try and make it slink off before them, and enable us in the end to get a fair chance at it in the open.

"Half an hour was spent in waiting for the spies to take up their positions, and then the whole body, chanting a hunting song so loudly that it could have been heard miles off, and must undoubtedly have broken the slumbers of the lion, marched up to the top, and spreading out, so as to take in all but the outskirts, where it was improbable that he would be, they entered the jungle shouting at the top of their voices, partly, no doubt, in obedience to my wishes, but quite as much to keep their own courage up. In this fashion, and amid cries of 'Get up! Get out, you dog! Where's the dog?' to which they trusted a good deal as likely to intimidate the lion, we passed right through to the other side, and though the ground had been beaten quite as well as it was possible for anything smaller than elephants to do, no vestige of the animal had been seen.

"Hardly, however, had the men begun to cluster out upon the open, before there was a shouting from the extreme left, which, when passed on through the stragglers, soon resolved itself into the lion having been seen there. Of course there was a general rush in that direction, which I accompanied, until I met a man who had come from the spot, and who said the brute had just showed itself and turned back. On hearing this I stopped those nearest to me and sent them to collect every one they could find, and in a few minutes two-thirds of the people had come around me. I then divided them into two bodies; the larger, led by all my hunters, except one, who remained with me, I sent to enter the jungle on the outer side and to beat through it, shouting and firing their guns; the other I took myself down to a stream which, at four or five hundred yards distance, fronted the spot where the lion had shown himself, and made them lie down in the bushes that lined it. About fifty men I stationed round the jungle, telling them never to cease making a noise, and I also removed the spies from in front of us.

"It took a long time to do this, and longer for the men to begin to beat, and we waited for an hour by the stream bank before anything happened. I had left my place and gone to drink, and as I turned to come back, a stir and rustle among the bushes where the men lay concealed made me think something must be in sight, and as soon as I got back, the man next me said, 'There he is!' and I caught sight of the lion standing under the shade of a solitary tree outside of the jungle, with his head turned in the direction of the beaters, evidently uncertain whether to await them where he was, or to take to flight. At last, doubtless considering that this was a different phase of the human character from the one he was accustomed to meet with during his midnight maraudings, he turned tail, and coming towards us in long easy bounds, was soon within a hundred yards of those concealed furthest down. Most fortunately I had told them all not to show themselves on any account before I did so myself, and so the brute, unsuspicious of danger, made for a ford near to which the hunter who had come down with me had stationed himself. At sixty yards he fired and rolled the animal over like a rabbit, it performing a complete somersault before it regained its legs; up the whole line jumped with a yell, and the lion, which I had first fancied was killed, continued his course the same as before, only, perhaps, rather stupefied by the shot, he abandoned the ford, and ran parallel to the stream, taking no notice of the people, many of whom shrank back as they saw him approaching their part of the line. I began to cover him when he was still two hundred yards off, and I think I kept the gun up too long, for when I fired at half that distance I missed clean. I made a better shot with my other barrel, rather too far forward, but just catching the point of the shoulder, and of course putting the limb *hors de combat*. *The brute appeared to be as cowardly by daylight as he was daring in the dark, for instead of charging he bolted under a small tree and lay down growling*, and in ten minutes all who were coming — and three-fourths of the men did so — had made their appearance, and were formed in a compact body behind me. He had not waited all this time very patiently; but when I fancied that I saw symptoms of his having a desire to slink away out of reach of the fast-arriving relatives of his victims, I had all the dogs set at him, and though only a few would go, and they could not have hampered his escape, yet they distracted his attention for a time.

"Our plan was a very simple one. The five hunters and myself were to walk up as close as we dared, and fire in volleys of three, and if we did not kill, and he charged, we were to bolt behind the natives for shelter. We walked up within thirty yards, and I and two hunters stood up while three knelt in front of us and fired, the lion growling furiously the while, but not attempting to move. The moment, however, the balls struck him — and with a lion crouched flat as he was, it was not to be expected that they could kill him unless one hit the centre of his forehead — he came straight at us, roaring horribly. My two companions, hardly going through the form of taking aim, pulled their triggers and joined those who had already fired. Fortunately the lion could not spring with a broken shoulder, and though he looked most unutterably savage, he did not get over the ground very fast, so I took a steady shot at the centre of his big chest, fully expecting to see him tumble over, but could not even see that it had struck him; and as he was getting very near I did not take a much better aim with the second barrel than the last two hunters had, and, like them, missed, turning as I did so, and running away for bare life. I was surprised to see how the men behind had diminished in numbers, but still there remained upwards of a hundred, who so far showed no sign of flinching, and I bolted in behind them and began to reload, altering my position when once the powder was down, so that I could see what was going on.

"The lion had charged up to within ten yards of them, and then, no doubt, awed, by their steadiness, he had pulled up, and was now walking slowly up and down like an officer in command, growling and showing his teeth, and looking a very noble animal with his heavy yellow mane floating around him. Very likely he would have remained like this until we had reloaded had not a young fellow in the first rank flung his assagai, with an insulting expression, at him; but as the spear-head entered he made two bounds forward, singling out the unfortunate man, who, however, met him pluckily, presenting him with his great six-foot shield to tear at, while he stuck him in the chest with his long and keen double-edged stabbing spear. As he did so there was a sudden jerk, as of a steel trap closing along the line, through which I was in time to catch sight of two more assagais being simultaneously plunged into the beast. All those who had run away hurried up, and a dense mass was formed, pushing and struggling to get into the centre, making the scene somewhat resemble a native foot-ball match I had once seen in the colonies. Such a contest could not possibly be continued long. Dozens of spears had been buried in the brute's body the instant it had reached the man, while, although I could tell by the shouting that they were still stabbing it, it was probably only a dead body on which they were wreaking their vengeance. Be that as it might, it was nearly half an hour before I could find an opening that led to the lion's carcass, and I do not think there was one solitary individual among all who were out that day who had not gratified himself by driving his spear into it; at any rate, its skin was a perfect sieve, and had at least five or six hundred holes in it. The price at which the victory was gained was comparatively small, only one man having received a fatal wound; while *the one upon whom the lion had sprung escaped with some severe gashes and a broken arm.*"

Those italics inserted in this narrative were not placed there by Drummond, but by the writer. They are intended to mark a propensity which he shared with many others to accuse the lion of cowardice while in the act of relating his deeds of desperation. This one it appears was "cowardly" because, with a shattered shoulder and other severe wounds, he did not at once attack a hundred armed men drawn up to receive him. Again and again had he penetrated into the midst of a populous village, and torn people out of their houses. All the same, he paused during the fight described, and was a poltroon. It is true that after walking up and down before his enemies like a lion of the Atlas as described by Gérard, he finally charged home and fought until cut to pieces. Still he was "cowardly." This is perplexing; there must be some standard by which courage is judged of in the case of lions that ordinary people know nothing about.

It is disappointing to find a man whom Lloyd calls "the well-informed Andersson," saying that "the length of a South African adult lion, from the nose to the extremity of the tail, is from eleven to twelve feet, . . . and his weight not less than from five to six hundred pounds." He knew all about the stretching of pegged-out skins, he had never seen a lion eleven feet long in his life, and yet he adds two feet, or at least eighteen inches, to the animal's average length, and a hundred pounds to its weight. Nine feet and a half is the average length of a well-known Indian tiger, which is certainly a larger animal than the lion, and both may occasionally reach a length of ten feet, but very rarely. Sometimes, also, lions weigh as much as five hundred pounds, although few persons have met with specimens so heavy; but beyond these measurements and weights, nothing is on record that deserves serious consideration. There is a perfect fog of contradictions about the animal's strength, leaping power, and his manner of carrying off prey; so that as far as testimony in these matters goes, no one can arrive at any conclusion. A lion stands about thirty-six inches high at the shoulder, and, of course, exceptional individuals may be taller. He can no more go straight with his head twisted over his shoulder than a man could; therefore, taking into consideration the length of his neck, those stories told about the manner in which lions bear off large animals in their mouths, and gallop away with oxen flung across their backs, have the disadvantage of being impossible. Thunberg asserts that one of these beasts will "attack an ox of the largest size, and very nimbly throw it over his shoulders, and leap a fence four feet high." Leveson says he leaps the stockade of a kraal whose palisades are six feet above the ground, with a steer in his jaws; and Sparman declares that he saw a lion carry off a heifer in his mouth, "as a cat would a rat." Drummond's lions sprang over thorn fences of an indefinite height, carrying their human victims; Gérard's made no difficulty about clearing the enclosures of Arab douars, while weighted with cattle. Montgomery Martin knew them to bear away horses and cows under like circumstances, and quite as many and as good authorities protest that all this is nonsense, and that they never did, and could not do, anything of the kind.

How much intellect this species possesses, and to what extent it can be cultivated, remains almost unknown. Their organization makes them subtle, fierce, and sometimes passionate beyond the limits of self-control, but they are, no doubt, capable of affection, and certainly exhibit marked preferences and dislikes. Apart from the instruction lions receive from their parents, — chiefly the mother, — and independently of anything which association may do for them, all are to a great degree self-taught; each one according to its capacity, to the extent of its opportunities, and correspondently with the character of its own mind. They design and carry out their conceptions, they imagine, and act the scenes suggested by fancy, they remember and combine their experiences.

Lions are not hunted with elephants in Africa. Dutch settlers in the southern part of this continent use horses, but only ride up within shooting distance, dismount, wheel their animals round so that they may receive the charge, if one is made, and then fire volleys with their roers — guns nearly as large as Asiatic and Mediæval wall-pieces. A number of other European sportsmen have also shot from the saddle; the advantage of this plan being that, in case the lion is only wounded, their horses will enable them to escape. Care is, however, necessary not to get too close; otherwise, so great is this beast's speed for a short distance, that a mounted man is almost certain to be overtaken.

The lion is a nocturnal animal, although in the more wild and desolate regions he may often be seen by day, especially in dark and stormy weather, and then either singly or in troops. Families of lions live together until the cubs are mature enough to shift for themselves; but a troop is a temporary co-operative association designed to drive game. Andersson states that he has seen "six or seven together, all of whom, so far as he could judge, were full-grown, or nearly so." Freeman relates that he once encountered a party consisting of ten lions. On another occasion he saw " five lions (two males and three females) in a party, and two of these were in the act of pulling down a splendid giraffe, the other three watching, close at hand, and with devouring looks, the deadly strife." Delgorgue once counted thirty lions formed in a hunting line. Many are really shot on foot in Africa, many more indeed than the tigers reported to have been killed in this manner in India.

Skaärm-shooting — the occupation by the hunter of a partially covered trench near a water-hole, — and the machan, or tree-platform, has also been adopted. Lions may often be seen walking about amid herds of antelopes on the African plains "like Caffre chieftains," as Delgorgue expresses it, "counting their flocks." The antelope knows that it cannot be caught so long as it keeps beyond the range of his first few lightning-like bounds, and thus its equanimity is in nowise disturbed by this destroyer's presence. Nothing but a stalk or an ambush will bring one of these fleet animals within their enemies' reach.

"Generally, however," says Andersson, "during the day a lion lies concealed on some mountain side, or beneath the shade of umbrageous trees or wide-spreading bushes. He is also partial to lofty reeds and long, rank yellow grass, such as occurs in low-lying 'vleys.' From these haunts he sallies forth when the sun goes down and commences his nightly prowl," and except the elephant and rhinoceros, there is no land animal in Africa that he cannot, and does not, kill. When lions attack the cattle of native rulers, their herdsmen, whose lives are held by native masters in no manner of account, are compelled to take their shields and spears and go after the marauder. There is no particular skill displayed save in tracking the beast to its lair, and the desperate close fighting which follows is due to the fact that the men know it is much better to be wounded or even killed, than trust themselves to the tender mercies of a negro chief who is enraged at the loss of his property. Namaqua Hottentots, who possess firearms, never take any risks. They go out in large parties, get into a safe place, and when a lion is provoked to charge, he is met with a storm of balls. A filthy little clay-colored Bushman will steal upon the sleeping beast with a caution and skill equal to its own. He has no weapon but a toy bow and tiny, often headless, arrow, poisoned with the entrails of the N'ga or Kalihari caterpillar, mixed probably with some form of Euphorbia. This savage wounds the sleeper without being himself seen, and an injury, however slight, is fatal. Delgorgue describes a lion-hunt by Caffres as follows :

"One of them, carrying a large shield of concave form, made of thick buffalo hide, approaches the animal boldly, and hurls at him an assagai or javelin. The lion bounds on the aggressor, but the man in the meanwhile has thrown himself flat on the ground, covered by his buckler. While the beast is trying the effect of his claws and teeth on the convex side of the shield, where they make no impression . . . the armed men surround him and pierce his body with numerous assagais, all of which he fancies he receives from the individual beneath the shield. Then these assailants retire, and the lion grows faint and soon falls beside the Caffre with the buckler, who takes care not to move until the terrible brute has ceased to exhibit any signs of life."

It is well known that, as a whole, the native populations of Africa display more enterprise and courage in the pursuit of dangerous wild beasts, than do those of Asia. But extraordinary and well-nigh incredible as are some of the stories about the temerity of certain tribes in lion-hunting as told by Freeman and Sir A. Alexander, the account given by Sir Samuel Baker ("Nile Tributaries of Abyssinia") of the Aggageers, or Arab sword-hunters of the Upper Nile, fully equals them. It is true that he did not see Taber or Abu Do, those slayers of elephants, cut a lion through the spine with their Solingen blades; but there is no doubt that these men encounter the animal on horseback and armed with their swords alone.

Brave as the Hamran Arabs were, and skilful, Baker, who has recorded their deeds, was not behind them in daring; and as the following narrative may almost be said to stand by itself in the records of hunting as an illustration of what can be done by a sportsman who is entirely courageous and cool, it is given in the words in which he has himself related his feat.

Some lions had been wandering about his camp for several nights, and they also gave him a good deal of annoyance by devouring game that he shot. "Under these circumstances," Sir Samuel says, "I resolved to circumvent one or the other of these beasts. On the following morning, therefore, I took Taber Noor, with Hadji Ali and Hassan, two of my trusty Tokrooris, and went to the spot where I had left the carcass of the buffalo I had killed on the preceding day. As I had expected, nothing remained, not even a bone; the ground was much trampled, and tracks of lions were upon the sand, but the body of the buffalo had been dragged into the thorny jungle. I was determined, if possible, to get a shot; and therefore followed carefully the trail left by the carcass, which formed a path in the withered grass. Unfortunately the lions had dragged the buffalo down wind, and, after I had arrived within the thick nabbuk and high grass, I came to the conclusion that my only chance would be to make a long circuit, and to creep up wind through the thorns until I should be advised by my nose of the position of the carcass, which would be by this time in a state of putrefaction, and the lions would most probably be with the body.

"Accordingly, I struck off to my left, and continuing straight forward for some hundred yards, again struck into the thick jungle, and came round to the wind. Success depended on extreme caution, therefore I advised my three men to keep close behind me with the spare rifles, and I carried my single-barrelled Beattie. This rifle was extremely accurate, and for that reason I chose it for this close work, when I expected to get a shot at the eye or the forehead of a lion crouching in the bush. Softly, and with difficulty, I crept forward, followed closely by my men, through the high withered grass beneath the dense green nabbuk bushes, peering through the thick covert with nerves strung to the full pitch and finger on the trigger, ready for any emergency. We had thus advanced for about half an hour, during which I frequently applied my nose to within a foot of the ground to catch the scent, when a sudden puff of wind brought the unmistakable smell of decomposing flesh. For a moment I halted, and looking round at my men, made a sign that we were near the carcass, and that they were to be ready with the rifles.

"Again I crept forward, bending and sometimes crawling beneath the thorns, to avoid the smallest noise. As I approached, the scent became stronger, until at length I felt that I must be close to the carcass. This was highly exciting. Fully prepared for a quick shot, I stealthily crept on. A tremendous roar in the dense thorns within a few feet of me suddenly brought the rifle to my shoulder;

K

almost at the same instant I saw the three-quarters figure of either a lion or a lioness within three yards of me, on the other side of the bush under which I had been creeping. The foliage concealed the head, but I could almost have touched the shoulder with my rifle. Much depended upon the bullet, and I fired exactly through the centre of the shoulder. Another tremendous roar, and a crash in the bushes, as the animal made a bound forward, was followed by another roar and a second lion took the exact position of the last, and stood wondering at the report of the rifle, and seeking for the cause of this intrusion. This was a grand lion with a shaggy mane; but I was unloaded. Keeping my eyes fixed upon the beast, I stretched my hand back for a spare rifle; the lion remained standing, but gazing up wind with his head raised, and snuffing in the air for the scent of an enemy.

"I looked back for an instant, and saw my Tokrooris faltering about five yards behind me. I looked daggers at them, gnashing my teeth, and shaking my fist. They saw the lion, and Taber Noor, snatching a rifle from Hadji Ali, was just about to bring it, when Hassan, ashamed, ran forward — the lion disappeared at the same moment. Never was such a fine chance lost through the indecision of gun-bearers. . . . But where was the first lion? Some remains of the buffalo lay upon my right, and I expected to find him most probably crouching in the thorns near us. Having reloaded, I took my Reilly No. 10 rifle, and listened attentively for a sound. Presently I heard within a low growl. Taber Noor drew his sword, and with his shield before him searched for the lion, while I crept forward towards the sound, which was repeated. A loud roar, accompanied by a rush in the jungle, showed us a glimpse of the lion as he bounded off within ten or twelve yards, but I had no chance to fire. Again the low growl was repeated, and upon quietly creeping towards the spot, I saw a splendid animal crouched upon the ground, among the withered and broken grass. The lioness lay dying from the bullet wound in her shoulder. Occasionally in her rage she bit her own paw violently, and then struck and clawed the ground. A pool of blood was by her side. She was about ten yards from us, and I instructed my men to throw a clod of earth at her (there were no stones), to prove whether she could rise, while I stood ready with the rifle. She merely replied with a dull roar, and I ended her misery with a ball through the head."

"Lions," says Andersson, "if captured when quite young, and treated with kindness, become readily domesticated, and greatly attached to their owners, whom they follow about like dogs." This statement is hardly worthy of its author, and the fact that these beasts are often kept in African villages, and made pets of by Asiatic rulers, does not at all warrant his sweeping assertion. He knew better than to suppose that a young wild beast did not inherit the traits of its ancestors, or that one cub was the same as another. Likewise there is no reason to doubt that he was acquainted with the incidents which constantly attend such experiments in the places mentioned. All this has already been discussed, but the lion's place in the opinions of those who live in the same land with him, and are unprepared to meet his majesty, is a more convincing proof with respect to his character than any other that could be advanced. A very small portion of mankind respect anything that they do not fear. Wherever lions exist under the conditions mentioned, they are dreaded, and with reason, and then, very often, their "daring and audacity almost exceed belief," according to Andersson, who after all expresses the sense of those writers in whose self-contradictory evidence they are called cowards. It was because men dreaded the lion that he became the emblem of wisdom in Assyrian sculpture and the type of courage in Hebrew poetry; that his head crowns the body of an Egyptian god, and that his form has been taken as a royal cognizance in the East and West. For no other cause is it that death is the penalty for any one but a ruler to wear his claws in Zululand, or that among the Algerian Arabs his whole body possesses magic virtues.

Lion flesh is eaten in various parts of the earth, although that counts for nothing with regard to its edibility, for men in certain phases of development eat everything. Andersson ate some ("The Okovango River") and found it white, juicy, and "not unlike veal." Much the same was said ages before his time in Philostratos' Life of Apollonius of Tyana, and though this work is doubtless an Alexandrian forgery, the evidence in this particular is just as good as if it were authentic.

In an account of this creature it remains to say a few words more about its intellect, and the conditions under which it is developed. Given the raw material of mind as a variable quantity in all beings belonging to the same group, the difference between them, apart from that which depends upon unequal endowment, results from the degree to which the exigencies of life force individuals to use that amount of intelligence which they possess. Existence to a lion is a very different thing in one place and another; it is difficult or easy, varied or monotonous, dangerous or safe, solitary or the reverse. In other words, those adjustments of internal to external coexistences and sequences which constitute what is essential in life, may be many and great, or few and small. In either case adaptations must be made, but unequal enlargements of faculty are the necessary results. Take, for example, the average lion and place him, as he is placed in fact, under the opposite conditions of having been born and reared in a desert, or brought forth amid a cluster of villages and trained to prey upon human beings. That such specimens cannot be the same needs no saying, and if not these, then not any who are differently placed; so that to go into some large province and write about this beast as if the few individuals met with summarized all the possibilities of its race, is manifestly absurd. Actually, and as far as he goes, a lion is as much an individual as a man; like men also, the more general resemblances and differences among them which are not due to organization, depend upon their position.

Diminish the quantity of game in the area where a lion lives, and its character is altered. Take away certain objects of prey, and replace them with others, and the brute will be more or less cunning, fierce, bold, enterprising, and active. He cannot live at all, without adapting himself to the character of those beings among whom his lot is cast, and as they change so will he change also. The same is true with respect to alterations in physical conditions.

Lions vary with sex; the lioness is usually less grave and inert, but quicker, more excitable, savage and enterprising than her mate. Once when Gérard was lying in wait by a dead horse a lioness arrived with her cub, but pretended not to see the hunter. She instantly pounced on her unsuspecting whelp, drove it out of harm's way, then made a detour, and stole silently back to kill him. This means maternal solicitude to the extent of temporary self-forgetfulness, presence of mind, rapid comprehension of the circumstances involved in an unexpected and unusual situation, determined purpose, and courage. Tigers constantly make false charges with the design of intimidating their foes; lions perhaps resort to this ruse less frequently, but they adopt other means to the same end. Much of their awe-inspiring appearance is due to causes acting independently of will; still, they deliberately attempt to excite terror. One night while Green and his friend Bonfield occupied a screen near a wateringplace, a lion passed and repassed, inspecting them closely. He wished the intruders away, but thought it imprudent to attack their position, and they objected to fire because the noise would frighten away elephants for which they were waiting. Then the lion walked off a little distance, lay down facing them, and reflected on the situation. Shortly he sprang up and began to cut extraordinary capers, at the same time setting up "the most hideous noise, neither a roar nor a growl, but something between the two."

The beast was trying to frighten off these unwelcome

visitors who might keep game at a distance and interfere with his supper. No one who watches young wild beasts, and more particularly those of the cat kind, can fail to notice that they continually rehearse the chief acts of their lives under the influence of imagination. A lion's memory is good, and he can be taught much. His judgment is excellent, and he seldom attempts what he is unable to carry out. In cold blood, prudence is one of his distinguishing characteristics, and he is also very suspicious and on the lookout for destructive devices and inventions of the only enemy he has reason to fear; that is to say, man. Thus, although parts of Africa may be said to be undermined with pitfalls, lions rarely fall into them and when this happens they often claw steps in their walls and get out. Not, however, out of the trenches dug inside of the fence round an Arab cattle pen, for there their enemies occupy its edge, and then it is seen that there are certainly occasions when lions meet inevitable death in a very dignified manner.

THE LEOPARD AND PANTHER

THOSE conflicting opinions we have thus far seen expressed upon the habits and characters of wild beasts, are not replaced by any unanimity upon the part of those who have described leopards and panthers. They have a less voluminous literature than the lion or elephant, but their temper and traits are disputed about in every particular, and even the place they occupy in nature.

The only difference between a panther and a leopard is one of size; or as G. P. Sanderson ("Thirteen Years among the Wild Beasts of India") expresses it, the distinction is the same as that existing between a "horse and a pony." Dr. Jerdon ("Mammals of India") states that they are merely "varieties of *Felis pardus*," and if the species-making mania were not so prevalent, one might wonder at men who constantly met with these creatures in Asia and Africa, and yet wrote about them as if they belonged to distinct groups, and had very little in common.

Major H. A. Leveson ("Sport in Many Lands") thus describes the panther: "This animal frequently measures eight feet in length from its nose to the end of its tail. It has a well-defined, bony ridge along the centre of its skull for the attachment of the muscles of the neck, which is not

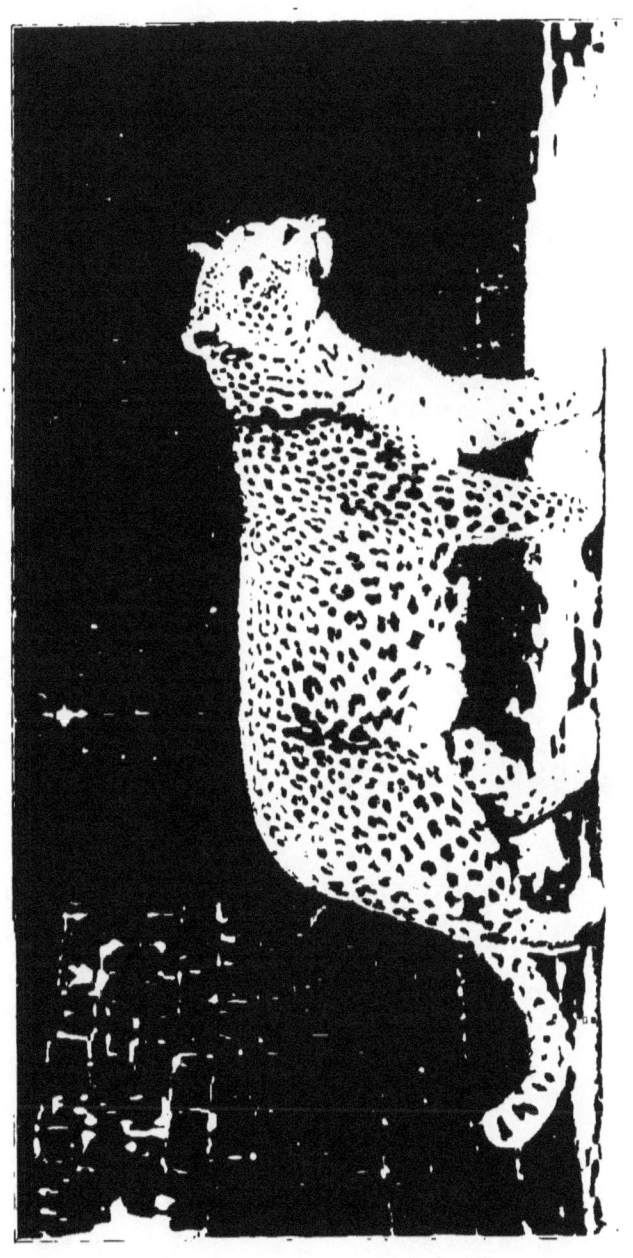

noticeable in the leopard or cheetah. The skin, which shines like silk, and is of a rich tawny or orange tan above and white underneath, is marked with seven rows of rosettes, each consisting of an assemblage of black spots, in the centre of which the tawny or fulvous ground of the coat shows distinctly through the black. Its extremities are marked with horseshoe-shaped or round black spots. Few animals can surpass the panther in point of beauty, and none in elegance or grace. His every motion is easy and flexible in the highest degree; he bounds among the rocks and woods with an agility truly surprising — now stealing along the ground with the silence of a snake, now crouching with his fore-paws extended and his head laid between them, while his chequered tail twitches impatiently, and his pale, gooseberry eyes glare mischievously upon his unsuspecting victim." Captain J. H. Baldwin ("Large and Small Game of Bengal") writes in much the same strain upon the specific differences between these varieties, and he is at a loss to understand how Dr. Jerdon and Mr. Blyth, Captain Hodgson and Sir Walter Elliot, can regard panthers and leopards as of the same species. The difference between their skulls — that of the leopard's being oval, while the panther's is round — is, he asserts, "of itself conclusive evidence upon this disputed question;" and besides that, "the two animals altogether differ from one another in size and character."

Technical discussions have been avoided so far as it was possible to do so, but here it seems necessary to say briefly that head-measurements as a basis for classification, whether among beasts or men, have always failed;

also that developed ridges and processes are for the most part merely concomitants of more massive skulls in larger animals whose muscles are of greater size; and that bulk by itself means very little, and varies in most cases very much. Finally, the coat-markings, in their minor details, of all animals whose skins are variegated, constantly differ in the same species. Among *Felidæ* one scarcely sees two lions with like manes, or two tigers with identical stripes. As for the spotted or rosetted groups, their spots not only vary in members of specific aggregates, but even upon different sides of the same creature's body.

Lockington ("Riverside Natural History") states that "the leopard (including both varieties of *Felis pardus* under this term) is very variable in size and color." Stanley, Emin Pasha (Dr. Schnitzer), and Hissman speak of those in Somali-land as much larger than any others in Africa, yet it is certain that there is but one true species now extant, and that this includes those forms already spoken of, together with the snow leopard of the Himalayas, the long-furred, ring-marked, bushy-tailed variety of Manchuria and Corea, and the "black tiger" of India and the Malasian Archipelago, which is nothing but a panther with its colors reversed, — a "sport," as G. A. R. Dawson ("Nilgiri Sporting Reminiscences") calls it, and which according to him is "of a uniform dull black color, with its spots (of a fulvous tint) showing in particular lights." Colonel A. C. McMaster proved that these dark cubs had been found in litters having the usual coloration. General Hamilton demonstrated the same thing, and Colonel Pollok ("Natural History Notes") states that "the black panther, which is very common towards Mergeri and Tavay, is only a *lusus naturæ*." He himself "saw a female panther near Shoaydung, with two cubs, one black and one spotted."

The "snow leopard" is very little known on account of the solitary and inaccessible regions it inhabits. "It is the rarest event," says Colonel F. Markham ("Shooting in the Himalayas"), "to see one, though it roams about apparently as much by day as by night. Even the shepherds who pass the whole of the summer months, year after year, in the area where it lives, that is to say, above the forests where there is little or no cover . . . seldom see one. . . . It is surprising and unaccountable how it eludes observation." He describes its ground color as being of a dingy white, with faint yellowish-brown markings, and represents the animal to be considerably smaller than its congeners of the hot country below. Captain Baldwin, however, saw a skin as large as a panther's. This was "of a light gray color, with irregular black spots. There was a black line running lengthways over the hind quarters, the hair was long on the neck, and the tail was remarkably long, ringed with black, and black at the tip."

An animal of the same species, and very like this, is confined to the equatorial belt of Africa. It is as rare as the "snow leopard," and has only been seen once or twice. Andersson ("Lake N'gami") reports that the "maned leopard" was mistaken by him for a lion. This name is a translation of the native title — N'gulula, and Leslie, who knew more about it than any one else, states that "a cub is gray, light, and furry. . . . The half-grown one, gray also, but the spots are faintly distinguishable. In the full-grown animal they are perfectly plain, but very dirty and undefined. There is also a peculiar gray hog mane." W. H. Drummond ("Large Game and Natural History of Southern Africa") also met with the N'gulula, and he, like Andersson, thought at first that it was a small lion, which it greatly resembled " in shape and color."

We may now turn from the varieties of *Felis pardus* and their external characteristics, to an investigation of those traits which have become organized in them during the long course of ages in which they have become specialized, physically and mentally, for a predatory life.

To know what an animal of this kind feeds on, and how it takes its prey, is also to know much about its structure, temper, and disposition. Neither lions nor tigers find the game upon which they subsist in trees, and the latter, therefore, rarely climb, while there is no account of the former having been seen to do so.

With the panther and leopard this is quite different. There are no climbers more expert than these beasts. As the Panama chief said to the explorer Oxenham, "Everything that has blood in it is food"; to these animals many things without blood, or at least without red blood, are food, for they eat the larva of insects, insects themselves, and birds' eggs; likewise many fowls, from the splendid peacock to a common crow, which, as Sir Samuel Baker remarks, "lives by his wits, and is one of the cleverest birds in creation." The panther preys on deer more commonly than any other kind of game, although it destroys reptiles, rodents, etc., and wild pigs in great numbers. Perhaps a wild boar, the "grim gray tusker" of Anglo-Indian tales and hunting songs, "laughs at a panther," as General Shakespear ("Wild Sports of India") declares. But all the weaker members of his race become victims to this spotted robber's partiality for pork. Monkeys, too, from the sacred Hanuman down through all secular grades, are eaten with avidity by these animals, and they kill great quantities of them despite their cunning. There is nothing alive of which monkeys are so much afraid.

Both leopards and panthers can endure thirst much better than tigers, and the latter are cave-dwellers to a greater extent than any of the larger *Felidæ*. They only drink once in twenty-four hours, and always at night. Their retreats lie amid low, arid, rocky hills covered with underbrush, traversed by gullies whose sides have been washed out into recesses by floods, and their rocks worn away into caves by weathering or percolation. They are much more active and energetic than their striped relatives, can better endure fatigue, and are, as a rule, bolder and more enterprising.

It is very far from being a fact, however, that "the habits of leopards are invariably the same"; that is an error into which Sir Samuel Baker was betrayed by the doctrine of instinct, and which has likewise been shared by nearly every other writer upon natural history. There is a certain sameness in the behavior of such creatures, as there is in that of all classes of animals leading similar lives; but this is as much as it is possible to say. In some localities, for example, the panther is strictly nocturnal; in others it appears that he hunts during the day nearly as much as at night. In no instance is he an organic machine. Far from it; this prowling marauder is the fiercest and most adventurous of wild beasts, astute to a degree, capable of using every faculty to its fullest extent, well able to take care of itself, and fatally skilful in compassing the destruction of others; a being in every way qualified to design and execute its projects, to achieve all those ends which courage and cunning enable it to attain, and quite fit to meet the ordinary emergencies that may arise during the perpetration of its acts of rapine and bloodshed.

The panther's cry — Gérard ("Journal des Chasseurs") calls it a "scream" — is often heard upon Indian hillsides when darkness begins to obscure the scene. Captain Baldwin describes it as a harsh, measured coughing sound, without much timbre or resonance, rather flat, in fact, and not at all like the roar of that animal it most resembles, — the American jaguar. Like most of the *Felidæ*, this species commonly gives tongue upon leaving its lair, or, at least, has been frequently reported as doing so. This is not a point of much moment, but it is a matter of considerable importance to the inhabitants of any village that may lie in the neighborhood, whether that ominous voice dies away in the forest, or appears to be approaching their dwellings. When a panther takes to man-eating, Colonel Pollok ("Sport in British Burmah") and Captain James Forsyth ("The Highlands of Central India") assert, "he is far worse than a tiger." Certainly, no records of such desperate ferocity exist in the case of any other creature of the cat kind; no other is reported to have taken like risks or to have succeeded in its fatal enterprises in the face of equal difficulties.

It is to be taken into consideration that a panther very rarely exceeds eight feet from tip to tip, or weighs more than a hundred and seventy pounds. Several writers have said that this animal's powers of offence are scarcely inferior to those of the tiger; nevertheless, nothing is more certain than the fact that with all its great strength, its exceeding activity, and formidable armature, a panther cannot stand before a tiger for a moment. It cannot overwhelm a man instantly, bite him through the body, or crush his life out with a single blow; and yet, unless like the superstitious people whom this fell beast destroys, we can imagine demons becoming incarnated to scourge humanity, nothing more terrible and deadly than a man-eater of this class can be conceived of. Captain Forsyth thus sketches a famous panther of the Seoni district, which he was in charge of when those scenes alluded to occurred. "This brute killed, incredible as it may seem, nearly a hundred people before he was shot by a shikári. He never ate the bodies, but merely lapped the blood from the throat. His plan was, either to steal into a house at night and strangle some sleeper on his bed, stifling any outcry with his deadly grip, or to climb into the high platforms on which watchers guard their fields from deer, etc., and drag his victim thence. He was not to be balked of his prey, and when driven off from one side of a village, would hasten round to the opposite side, and secure another person in the confusion. A few moments accomplished his murderous work, and such was the devilish cunning he joined to his extraordinary boldness, that all attempts to find and shoot him were for many months unsuccessful. European sportsmen who went out, after hunting him in vain, would often find his tracks close to their tent doors in the morning."

It is about time that the usual explanation given for this kind of exceptional conduct upon the part of a beast of prey by those writers who think it necessary to allude to their character, otherwise than in general terms, was banished from descriptive natural history. The course of thought upon the natural relations which subsist between men and brutes, seems to run somewhat in this wise. At sometime, somewhere, and somehow, all inferior denizens of this earth were made to appreciate and fear human superiority. That impression was transmitted as an instinct, and is in full force now. When, therefore, a predatory animal does such violence to its nature as to eat a man, the shock, which according to conventional ideas always attends great crime, unhinges its mind. A kind of madness ensues. It becomes wild, and is driven by Furies like an ancient Greek guilty of sacrilege, or early Christians who, as reported by Gregory the Great and many others, had swallowed devils. Instantaneous change of character is the consequence, and the creature henceforth thinks, feels, and conducts itself in a new and terrible manner.

That is about the sum and substance of most statements bearing upon this subject, and there is not the slightest foundation in fact for any of them. This question has been considered in the abstract; but with regard to the panther's character the truth is that, in the way stated, no respect for mankind is discoverable in his conduct. It is indeed notoriously otherwise; and this is nowhere more clearly shown than in the records of observations made by men who were convinced that all species of wild beasts instinctively feared them. "The Old Shekarry" (Major Leveson) writes ("Hunting-Grounds of the Old World") to this effect: "Panthers, like all forest creatures . . . are afraid of man, never voluntarily intruding upon his presence, and invariably beating a retreat if they can do so unmolested." Then this authority goes on to tell what he has learned about panthers in the course of an experience rarely equalled for extent and variety. They are "more courageous than the tiger. . . . The panther often attacks men without provocation." When one "takes to cattle-lifting or man-eating he is a more terrible scourge than a tiger, insomuch as he is more daring and cunning." He relates how this timid creature that never voluntarily obtrudes himself upon human presence, fights hunters on all occasions; how the beast broke into his own camps, carried off dogs that were tied to his tent pole, and much more to the same effect.

There is no difficulty in finding exploits of the same kind; Rice, Inglis, Forsyth, Barras, Shakespear, Pollok, Baker, Colonel Walter Campbell, who saw the man riding next him in a party of horsemen, torn out of his saddle, or Colonel Davidson moving with a column of troops around whose encampments the sentinels had to be doubled to prevent panthers from killing them, all tell the same story.

L.

"The tiger is an abject coward," and so is the lion. Panthers are audacious, but they run away upon instinct, like Falstaff. No qualifications, no reservations, are made, no middle ground is taken, only a series of facts is given, which prove, so far as anything in this connection can be said to be proved, the incorrectness of what was insisted upon in the first place.

The opinion that a wild beast that has tasted human blood is thereby metamorphosed morally, "undergoes a transformation of emotional psychology," as Professor Romanes expresses it, scarcely deserves a serious refutation. There is not the slightest reason why any such change of character should take place, and of course it does not. But the fact of a wild beast's taking to man-eating is a sufficient cause for an alteration in habit. What modifies the animal then, however, is not the fact of killing a man, but the discovery of the ease with which he can be destroyed. Under these circumstances the brute simply substitutes one kind of game for another; it becomes used to the feeble attempts at opposition met with, and goes on with its murders. Where the state of affairs is different, where people are ready to combine against such scourges, to anticipate their designs, pursue, circumvent, and slay them, these beasts of prey do not devour men; they keep as far from them as possible.

It is doubtful if it could be shown that panthers are more prone to anthropophagous habits than other brutes, but the evidence is strongly in favor of the fact that they fight human beings more readily. Their ferocity and hardihood are exceptional among the *Felidæ*.

The Leopard and Panther

The panther described by Forsyth set at naught quite a number of favorite theories. His conduct was indeed very different from that which might have been expected if the main features of character common to his family are like those which are said to exist. The relations of cause and effect were not set aside for his benefit, and therefore, instead of being at once prepared to do the things he is known to have accomplished, there must have been some period of preparation. Of all things it is the most improbable that this animal set out on an expedition at haphazard. Perception, foresight, comprehension, judgment, resource, were not suddenly conferred upon him when he arrived at his destination and taken away when he left. He must have added observation and training to his innate qualities. How easily or to what extent this was done we cannot decide; for to imagine that a wild beast could come out of the forest, and instantly become an experienced master of an entirely new set of circumstances and have the ability to take advantage of every opportunity and overcome all opposition, is preposterous; is nothing less than to suppose an effect without a cause. The brute in question gave terribly convincing proofs that it understood the situation in its entirety, and how this could have been the case unless it was known, in what way known without being learned, and how learned without a mind passing through ordinary processes, does not appear.

To isolate the traits of an animal and consider them separately is a mistake. It is to fall into the same error that Stallo and the transcendental school in physics have made with reference to the attributes of matter. These abstractions of the mind are not identical with realities in nature. They cannot be studied by themselves without distorting the subject to be represented. Compared with that of other great cats the panther's conduct shows that he is braver than the rest. But this is only an empirical conclusion and throws little light upon the animal's character. We are not in a position, however, to analyze this in such a way as to show the relative development of its traits, or to say how far excess in one direction alters the general disposition.

So far as the brute's behavior goes, the following narrative will be found to bear upon several points that have been discussed. Colonel Barras ("India and Tiger Hunting") had pitched his camp in the Murree jungles, and it was crowded with the usual supernumerary attendants, together with elephants, gharry bullocks, horses, and dogs. One night as he and his companions — Messrs. Sandford and Franks — lay upon their camp beds in the deep slumber that follows a hard day's work, they were awakened by "a furious roaring." It appears that a panther had come among them, and seized upon a pet dog belonging to the Colonel then tied to his tent pole.

The brute, finding that it was impossible to carry off his prey, became enraged. Everybody turned out, and the panther made off in the midst of the hubbub. But his visit was looked upon as a challenge, and they resolved to postpone any further proceedings against tigers in that vicinity, until this marauder had been hunted. Orders to that effect were issued to the head shikári, and that worthy acted upon them with such success as to report next morning that the trackers had marked him down. "After the usual hot march of three or four miles," says Colonel Barras, "we came upon the chief shikári, who was speedily to place us face to face with our hidden foe. On arriving at the scene of action, we found that the panther had taken up his quarters on a steep hillside which was much more thickly covered with cactus plant than usual. The top of the hill was flat . . . and devoid of cover. The last short rise up this eminence was so steep that a line of beaters had drawn themselves up in tolerable safety all along the crest, prepared to hurl showers of rocks and stones down the declivity, should the panther take an upward course. All of them, however, then maintained an immovable attitude and a profound silence, whilst in a whisper scarcely to be heard, our guide pointed out the exact bush in which the enemy was said to be concealed. We divided the distance around it, and gradually closed in towards the centre of attraction, till not more than five or six yards separated us from the place. . . . Here we paused in circumspection; no sound struck upon the ear, nor did so much as a leaf quiver a warning to the eye. But though invisible to us, we felt that the animal was aware of our presence, and that its eyes were fixed upon us as it crouched for a spring."

Still the panther remained quiet, "and whilst the party were discussing various projects, my dog keeper asked permission to ascend the slope of the amphitheatre on which we were standing, so that he might join the line of beaters on the ridge above. Permission was given, but he was strictly enjoined to make a circuit round the tract of bushes, to enter which would have been dangerous. He had not gone many yards, however, when with true native perversity he struck well into the middle of the cover, and stumbled right upon the panther, which to his no small dismay sprang from a bush only a few feet in front of him. . . . The brute suddenly appeared before us, going at a great rate through the underbrush. As it flashed across a small open space we all took snap shots, none of which took effect, and the animal dashed into a deep ravine and disappeared." Nothing now remained except to drive the game; that is to say, post the guns at a point where the beast would most probably attempt to break out, and cause the beaters to advance towards it. This was done, the signal given, and "the perfect stillness was instantly replaced by a wild shrieking, the rushing sound of falling rocks, and a waving about of people and herbage as though the whole mountain were about to slide into the valley beneath. No panther could resist such a pressing invitation to move as this was, and our friend accordingly started off at full gallop for other quarters," which he again reached without being hit, and presently the report came that the game had taken refuge in a dense clump of cactus on top of the hill. While messengers were despatched for rockets to drive it out, the party agreed to take lunch, and the "tiffin basket" was placed on the shady side of that impenetrable cover where the panther lay.

"For a few moments," continues Colonel Barras, "we sat quite still. Then it occurred to us to try and peep through into the centre of the mass of cactus to see if we could make out the whereabouts of its present occupant. . . . Not seeing anything, our thoughts reverted naturally to the basket. There it stood, just on the other side of Sandford. I stretched across him to reach it with my right hand, and had just grasped the handle, when a succession of short, savage roars broke upon my ears, mingled with the wild shouts of the natives, who were evidently being chased by the ferocious brute. At this time I felt that my hat would probably do more for me than my gun, so I crushed the former down on my head, seized the latter, and faced the enemy. The panther meanwhile had floored a beater and got him by the arm, but dropping him at once, came at me with lightning bounds. Owing to the beast's tremendous speed, I could see nothing but a shadowy-looking form, with two large, round, bright eyes fixed upon me with an unmeaning stare as it literally flew towards me. Such was the vision of a moment. . . . I raised my gun and fired with all the care I could at such short notice, but I missed, and the panther bounded, light as a feather, with its arms around my shoulders. Thus we stood for a few seconds, and I distinctly felt the animal sniffing for my throat. Mechanically I turned my head so as to keep the thick-wadded cape of my helmet in front of the creature's muzzle ; but I could hear and feel plainly the rapid yet cautious efforts it was making to find an opening so as to tear the great vessels that lie in the neck. I had no other weapon but my gun, which was useless while the animal was closely embracing me, so I stood perfectly still, well knowing that Sandford would liberate me if it were possible to do so. . . . As may be supposed, the panther did not spend much time investigating the nature of a wadded hat-cover, and before my friend could get round, and fire without jeopardizing my life, the beast pounced upon my left elbow, taking a piece out, and then buried its long, sharp fangs in the joint till they met. At the same time I was hurled to the earth with such violence that I knew not how I got there, or what had become of my gun. I was lying on the ground with the panther on top of me, and could feel my elbow joint wobbling in and out as the beast ground its jaws with a movement imperceptible to the bystanders, but which felt to me as if I were being violently shaken all over. Now I listened anxiously for the sound of Sandford's rifle, which I knew would be heard immediately, and carefully refrained from making the slightest sound or movement, lest his aim should be disturbed. In a few seconds the loud and welcome detonation, which from its proximity almost deafened me, struck upon my ear, and I sat up. I was free, and the panther had gone " — bounded away shot through the body with a heavy rifle ball, into an acacia and karinda thicket, from which it had to be driven by rockets.

"Just as the interior of the thicket became lighted up, and the crackling of the herbage was at its loudest, the animal roused to frenzy, by the overwhelming character of the attack, girded itself up for a last desperate effort. . . . It rushed from its now untenable hiding-place, swift and straight as an arrow upon Sandford and myself. He fired both barrels at the beast without stopping it in the least." The Colonel, whose wounded arm had been bound up, now carried a hog spear. "We had only time," he says, "to open out one pace from each other, and the momentum with which the animal was coming, almost carried it past us. As it brushed my right leg, however, I saw it twist its supple neck, and literally stop itself by clasping Sandford's thigh in its extended jaws, bearing him to the ground, where they lay for a moment in a close embrace. I at once adjusted my spear behind the animal's shoulder, and with a steady thrust drove it straight through the heart. Franks fired at the same instant, and it would be difficult to say which of us caused the panther to give up his last breath. It was dead though, yet it still retained the position it had in life, and its teeth were so firmly locked in the flesh of its foe, that I could not open the jaws with one hand — they felt like iron to the touch."

There are a number of narratives of like import with this, but neither in these, nor in the accounts we have of conflicts with other wild beasts, has anything been said concerning the principle upon which they fight. Briefly, no brute deliberately engages in conflict without thinking that the advantage is altogether on its own side. They may be, and often are, mistaken, but brutes never fight fairly with intention. Only man does that, civilized not savage man, whose motives are such as other creatures know nothing about.

Inglis ("Work and Sport on the Nepaul Frontier") relates an experience of his own with a leopard — it may as like as not have been what is here called a panther — that includes a good many points which have been touched upon,

—the much talked of eye power, this brute's instinctive avoidance of man, etc.,—and it is therefore inserted by way of illustration.

"I was camped out at the village of Purimdaha, on the edge of a gloomy Sal forest, which was reported to contain numerous leopards. The villagers were a mixed lot of low-caste Hindus and Nepaulese settlers. They had been fighting with the factory, and would not pay up their rents, and I was trying, with every prospect of success, to make an amicable arrangement with them. . . . It was the middle of April. The heat was intense. The whole atmosphere had that coppery look that betokens extreme heat, and the air was loaded with a fine, yellow dust which the west wind bore on its fever-laden wings, to disturb the lungs and temper of all good Christians. The *Kanats*, or canvas walls of the tent, had all been taken down for the sake of coolness, and my camp bed lay in one corner, open all round to the outside air, and only sheltered from the dew. It had been a busy day. I had been going over accounts, and talking with the villagers until I was hoarse.

"After a light dinner I lay down on my bed, but it was too close and too hot to sleep. By and by the various sounds died out. The tom-toming ceased in the village. My servants suspended their low-muttered gossip around the cook's fire, wrapped themselves in their white cloths, and dropped into slumber. Toby, Nettle, Whiskey, Pincher, and the other terriers looked like so many curled-up hairy balls, and were in the land of dreams. Occasionally a horned owl would give a melancholy hoot from the forest, or a screech owl raise a momentary and damnable din. At intervals the tinkle of a cow-bell sounded faintly in the distance. I tossed restlessly, thinking of various things, till I must have sunk into an uneasy, fitful sleep. I know not how long I had been dozing, when of a sudden I felt myself wide awake, but with my eyes yet tightly closed.

"I was conscious of some terrible, unknown, impending danger. I had experienced the same thing before when waking from a nightmare, but I knew that the peril was now real. I felt a sinking horror, a terrible and nameless dread, and for the life of me I could not move hand or foot. I was lying on my side and could hear distinctly the thumpings of my own heart. A cold sweat broke out behind my ears, and over my neck and chest. I could analyze every feeling, and knew there was some *Presence* in the tent, and that I was in instant and imminent danger. Suddenly in the distance a pariah dog gave a prolonged melancholy howl. As if this had broken the spell that bound me, I opened my eyes, and within ten inches of my face there stood a handsome leopardess gazing steadily at me. Our eyes met, and how long we confronted each other I know not. It must have been for some moments. Her eyes contracted and expanded, the pupils elongated, and then opened out into a lustrous globe. I could see the lithe tail oscillating at its extreme tip with a gentle waving motion, like that of a cat when hunting birds in a garden.

"Just then there was a movement among the horses. The leopard slowly turned her head, and I grasped the revolver that lay under my pillow. The beautiful spotted monster turned her head for an instant, showed her teeth, and then with one bound went through the open side of the tent. I fired two shots, which were answered with a roar. The din that followed would have frightened the devil. It was a beautiful, clear night with a moon at the full, and everything showed as plainly as at noonday. My servants uttered exclamations of terror. The terriers went into an agony of yelps and barks. The horses snorted and tried to break loose, and my chowkeydar, who had been asleep on his watch, thinking a band of Dacoits had come upon us, began to lay about him with his staff, and shout, '*Chor! Chor! lagga! lagga! lagga!*' that is, 'thief! thief! lay on! lay on! lay on!'

"The leopard was hit, and was evidently in a terrible temper. She halted not thirty paces from the tent, beside a Shanum tree, and seemed undecided whether to go on or return and wreak her vengeance on me. That moment of hesitation decided her fate. I snatched down my Express rifle, which was hanging in two loops above my bed, and shot her right through the heart.

"I never understood how she could have made her way past dogs, servants, horses, and watchman, into the tent, without raising some alarm."

Thus far, whether in courage, enterprise, and skill, whether in sagacity, or desperation of attack and defence, nothing has been found to traverse W. H. Lockington's opinion ("Riverside Natural History") to the effect that panthers, "relatively to their size, are the fiercest, strongest, and most terrible of beasts."

In ancient Egypt and modern Abyssinia lions formed part of the royal paraphernalia. Marabouts lead around sacred animals of this species in North Africa, and if they occasionally kill somebody, the public in those parts understand that he was a sinner who deserved his fate. Leashed tigers also were not uncommon in the courts of Hindu rajahs, but since the time of the Indian Bacchus, whose car they drew, panthers have rarely appeared in parades. These savage brutes do not lend themselves to peaceful pageants. From all accounts they are the most intractable and untrustworthy of creatures — the least susceptible of instruction, says Sanderson ("Thirteen Years among the Wild Beasts of India").

Panthers have often been seen associated in families, but they do not display what Professor Romanes calls "the collective instinct in hunting." They can supply their needs without resorting to these manœuvres, and therefore have not formed the habit of practising them.

It sometimes happens that *Felis pardus* in all its forms has to give up spoil. The lion takes its prey away, and so does the tiger. Occasionally some blundering, black rhinoceros comes upon the scene and puts the panther to flight, or a herd of wild hogs does the same. *Kuon rutilans*, the wild dog, is reported to be in the habit of appropriating their supplies, and J. Moray Brown ("Shikar Sketches") states that he had personal knowledge of this fact. Upon the whole, however, the beast in question is not much molested.

Over-boldness is disadvantageous to any animal, and panthers suffer from their temerity in the way of getting trapped more frequently than other members of their family. General Morgan ("Memoirs") remarks that "it is a very common thing to catch a panther," but nobody has said the same of other *Felidæ*. The difficulty lies in comparing these species so as to assign the phenomenon to its real cause. The question is, how does it happen that a panther walks into a pit more frequently than a tiger? It cannot be said that it is because the latter has the more intelligence; facts do not sustain such an explanation, and yet the absence of deliberation stands in a direct relation with incompleteness of mental development.

It might be argued that the dissimilarity was due to temperament, and that while neither could be absorbed by one idea — that of committing a murder, for instance — without some temporary disregard of everything else, the panther was more liable to this state of mind than its relative. In ordinary parlance such a tendency would be called courage, and its opposite timidity, although that is rather a loose manner of speaking. However the truth may be, there is no doubt that a tiger will often come up to a bait fixed over a pitfall, examine it carefully on every side, and finally walk off with that pleasant grin of his, while the panther precipitates himself into the cavity.

This beast is very partial to dog meat, and the canine population of countries where panthers abound have an abiding fear of them. Sir Samuel Baker ("The Rifle and Hound in Ceylon") says that his dog "Smut," who weighed a hundred and thirty pounds, and was "a cross between a Manilla bloodhound and some big bitch at the Cape," made a practice of hunting leopards on his own account. This was a very unusual thing, however, since the largest breeds of the East, Poligar dogs and Tibetan mastiffs, would certainly be at a great disadvantage in such an encounter.

While the latter was encamped upon the Settite River, an Abyssinian tributary of the White Nile, one of these animals sprang into the midst of a circle of men resting around a watch fire and carried off a dog. To invade a hunters' camp on this kind of an errand is quite common with the panther, and many exploits of his under such circumstances have been put on record. In India the villanous pariahs that swarm in every village are his constant victims. If one of them goes into the jungle, there is apt to be a momentary scuffle in the dry grass, a stifled yelp, and the dog vanishes. So in rice fields and around cattle camps where the Gwallas build their temporary residences. Principally, however, the panther gets game of this kind from permanent settlements infested with "curs of low degree." Panthers know them well, and act accordingly. During the night one approaches the outskirts of a village and so far reveals his presence as to show the dogs, who are always prowling about, that some strange animal is near. Now they in turn are well aware of the tricks that panthers play, but on the other hand can by no means resist their ingrained propensity to make a display of courage, which they probably possess in a less degree than any carnivora. As soon as these pariahs discover something that conceals itself, the idea which naturally takes possession of their minds is that this cautious conduct is due to a fear of themselves. The pack instantly darts forward, and stops. These brutes endeavor to get self-encouragement out of absurd antics; they leap, they howl, they ramp and rave, until one of them, more excitable than the rest, so far forgets itself as to approach the intruder too closely. A shadowy form bounds upon it, and all is over.

If panthers were contented to kill these animals only, no reasonable objection to their deeds could be made. Unfortunately this is not the case; sheep, goats, pigs, horses, cattle, and their owners, all are destroyed; and when some casualty more exasperating or tragic than usual occurs, public opprobrium descends upon the hereditary huntsman of the commune with true Asiatic violence and unreason. Is he, the accursed, supported in case and affluence in order to snore like a swine while people and their property are thus devoured? Oh Ram! Ram! Ram! May the choicest curses light upon him, may he be beset by all devils whatsoever! Then the official, who is wholly blameless, and except by accident cannot hope to do anything against a beast like this, curses the panther, his fate, his fellow-citizens, and himself; after which everybody forgets the matter.

No prudential reflections interfere with a panther's singleness of purpose when on the prowl. Blood is his object, and blood he intends to have, so the upshot is that he often finds himself at the bottom of a pit shaped like an inverted pyramid that it is impossible to dig out of. What subsequently happens depends upon the demand for wild beasts. If an agent of Jamrach's has left an order for panthers, or some native ruler signified his will that they be forthcoming immediately, the captive's life is safe. Men arrive in the morning with something that looks like a magnified crate. It is inverted over the pit's mouth, earth is thrown in, the floor rises and with it the captive, until the animal is forced into this temporary cage. Bamboo crosspieces are then slipped under and secured, and very shortly he is *en route*. If the destination be a zoölogical park or menagerie, it is said that the beast will live longer and develop physically more completely in captivity than it would in a state of freedom. This is, to say the least, doubtful. Much might be advanced upon the subject, but biological discussions are out of place here, and it is enough to point out the fact that this opinion must be purely arbitrary, since no vital statistics exist from which such a conclusion could be legitimately drawn.

Returning to the subject of traps, they are not always constructed alike. Besides excavations there are enclosures that must be entered intentionally or not at all. These are made by driving palisades deep in the earth, roofing them, and cutting a sliding door in the side. It is connected with the bait by a string in such a manner as to drop when this is touched. Tigers are seldom taken in by these inventions, but the panther is frequently caught, especially if a live animal be placed in the trap. How he reasons upon the unusual circumstances then presented we do not know. Perhaps there is little or no deliberation upon what he ought to do, and the brute merely acts in obedience to its immediate impulses. But if we examine the behavior of panthers that go into villages to kill men, in all instances of this kind the animal's conduct is marked by a union of skill and daring with cunning and circumspection. What makes him lose his prudence in face of a trap? Except himself, there is not a great cat in Asia that would not be apt to see into this device and keep out of danger. The panther, however, enters the enclosure. Such appears to be a fair statement of facts relating to the brute's character and habits in this connection, but no attempt is made towards explaining them.

In certain parts of India panthers are netted. That is, nets are hung about ten feet high behind which the hunters stand with spears. It is not *jouer de rigueur* to use rifles unless these defences are leaped.

In the event of the barrier being bounded over, the result to the huntsmen depends greatly upon the way in which the beast attacks. Some animals of this species have a curious habit, under such circumstances, of trying to kill all their enemies at once. Much the same has been said of tigers. Sir J. E. Alexander ("Expedition into Africa") speaks of the spotted cats of that country as flying about among a crowd of enemies, striking first at one and then at another. In such a skirmish nobody might be seriously injured. On the other hand, they cannot be counted upon to act in this manner, and if, like Barras' panther, one singled out a particular man and fastened upon him, nothing, it is likely, could save his life except prompt interference upon the part of his companions.

With regard to its attack upon game, the mode in which this animal takes its prey has been definitely settled in several different ways, as is the case also in respect to the manner in which its prey is eaten. Colonel Pollok ("Sport in British Burmah") remarks that "there is a peculiar and singular distinction, with regard to the mode of breaking up their prey, between the tiger and the panther, the latter invariably commencing upon the fore quarters or chest." General Shakespear, nevertheless, came unexpectedly upon a panther that had just killed a cow in the Bootinaut correa, and it was feeding upon one of the hind quarters, "a large piece of which had already been consumed." Colonel Barras and Captain Forsyth consider the throat to be the part first fastened upon, Baker states that the body is at once torn open to get at the viscera, and Inglis, Leveson, and others explain that panthers suck the blood of their victims before anything else.

Similar dogmatic opinions and exclusive views of the way in which a panther or leopard kills game have been advanced. They are said to break the neck with a blow of their forearm; and also never to do so, not being able in the case of large animals, and with small ones this being unnecessary. Some authorities maintain that the cervical vertebræ are dislocated by twisting the head; others, that the head is bent backward till the neck breaks.

Hon. W. H. Drummond ("Large Game and Natural History of South and Southeast Africa") says that "leopards and panthers are very numerous in that country." He likewise apparently regards these varieties as distinct species, and writes about the "*ingwe*" or *Felis leopardus*, the "*N'gulula*" or maned leopard, and *Felis pardus*, the true panther, as if two of these, at least, belonged to different groups.

Strangely enough to any one acquainted with the characteristics of the Asiatic panther, Drummond asserts that the leopard, which is a comparatively rare animal, is, although of smaller size, the fiercer and more dangerous of the two. He explains that its rarity is more apparent than real, and depends upon the creature's "nocturnal habits and the thickness of the jungles they lie in," so that sportsmen only "occasionally come across them by accident."

It is singular, however, that a hunter who had passed a number of years in a country where they abound, should have been so little impressed by the prowess of a beast which, at least in Central and West Africa, is very destructive to human life. It must be the case that the brute's character varies somewhat with locality, yet Drummond's narrative portrays a condition of things under which its native ferocity and aggressive nature should have been developed and not diminished. However this may be, the pale, almost white-skinned panther, whose light color is very conspicuous in its rosette, was plainly regarded by Drummond as a much less formidable foe than its congener of the Indian jungles, or even than its relations which Baker and others found in the northern parts of Africa.

Still, he admits that "common leopards, *i.e.* the two forms locally known under the name of *ingwe*, are much to be dreaded when brought to bay, and that anecdotes innumerable might be related of instances where they have killed or seriously injured both white and black hunters. The virus of their bite is very great. I remember once seeing seven men belonging to a Zulu village awfully torn and mangled by a single animal, and the wounds remained open for a long time, and ultimately left great scars. On the other hand, I know of several who have died where the injuries received were not such as to have been generally fatal." Sir W. C. Harris mentions it as a peculiarity of the leopard's attack that it strikes at the face; Drummond says nothing about this trait, and the former author probably fell into some confusion of ideas, caused by the well-known tendency of this species to tear open the great vessels of the throat.

Panthers and leopards are only varieties of the same species, yet while the reputation of the former is such as has been stated, hunters often speak of the latter as if it were nearly harmless so far as human beings are concerned. Leopards are described as having been shot right and left in the jungle, treed by dogs and killed on limbs without difficulty, pelted from the doorways of deserted huts, and speared in the open from the saddle. Leveson, Drummond, and Baker relate experiences of this kind, but the literature of the subject contains many very different accounts of their prowess. Both in Asia and Africa they have often been found to be extremely dangerous and destructive animals. There is good reason why in heraldic blazonry the leopard should be represented as *passant gardant*. The designers did not know it, but the fact is that no animal capable of doing so much harm, and that has as many evil deeds to answer for, is at once so enterprising, so stealthy, and so full of cunning. Compared with him, the greater *Felidæ*, on the one hand, and that much-abused assassin and robber, the fox, upon the other, are "mild-mannered," and might be called bunglers.

When a tiger — and the same may be said of the lion — attempts to carry out a scheme he has formed for the surprise and murder of some man whose whereabouts he has ascertained, the design is often more complete than the execution. His heavy yet muffled tread is sometimes heard, he breaks dry sticks, rustles as he moves through parched herbage, waves long grass in passing, so that any experienced eye can tell he is there, puts his head out of cover prematurely, is apt to cross open spaces when a circuit ought to be made; again, he cannot keep his tail still, and as the moment approaches for making an end of his victim, anticipation of the pleasure of putting the man to death and devouring him overcomes his caution, and he begins to purr. This is not a loud sound, but it is a very impressive one, and has been frequently heard. But no creature's senses can give warning of a panther's or leopard's approach. Few people ever heard or saw one of these beasts while coming. They steal upon their prey with the silence and certainty of death. Their stalk is the perfection of skill. The attack is rapid and fierce beyond comparison; and afterwards, unless the ground is such as will retain a trail, this animal cannot be followed. It is the most difficult to mark down of all beasts of prey, the hardest to track on account of its many tricks. No kind of game is so often hunted unsuccessfully.

Leopards get the advantage over a being far cleverer than any other forest animal. Monkeys of all species detest tigers, but have an intense dread of the spotted cats. They "swear" at the former, but fly from the latter, and as for men, monkeys deride them. Panthers and leopards catch these creatures in trees, on the ground, by day and by night; while they are on the alert, and in moments when an apparent absence of danger lulls these astute little beasts into a fatal feeling of security.

A cattle-lifting panther, according to Pollok and Forsyth, is more destructive than a tiger. On the great ranges where herds graze during the time when pasture is destroyed by drought in a good part of India, the depredations of these beasts cost the owners dearly, and they likewise take a constant toll from those animals, cows principally, which are kept at villages. A buffalo under ordinary circumstances is safe, even if alone; and when the herd is united to resist, even he with the stripes has not the slightest chance of success.

Cows, however, are the especial prey of panthers. In India these are of comparatively small size, and preternaturally imbecile. The *Bovidæ* are not a gifted family at their best, and when domestication relieves them to a great extent from the necessity of taking care of themselves, they lose much of the faculty which in wild forms is developed under the stress of necessity. Year after year, and age after age, the panther has been murdering Indian cattle in the same way, and they have never originated the slightest measure of precaution or defence. The full measure of their weakness of mind has been taken by the enemy, and when he concludes to give up hunting, except as a pastime, and live on beef, his prey may be said to come to him.

In 1863 Captain Forsyth hunted panthers on the higher Narbadá, under the auspices of an old shikári, an unspeakable scoundrel, who had killed more of them than anybody else whose exploits the annals of sport with large game perpetuate. Bamanjee (the Brahman) seems to have been exceptionally honest in his dealings with the Captain, and to have given him an opportunity, rarely accorded to the hunters whom he swindled, for making observations upon the habits and character of these beasts. Forsyth relates his experiences in a way that will serve as a summary of what has been already said about *Felis pardus*. "The number of these animals in the districts around Jubbulpúr is very great. The low rocky hills, . . . full of hollows and caverns, and overgrown with dense scrubby cover, afford them their favorite retreats; while numbers of antelope and hog deer, goats, sheep, pariah dogs, and pigs supply them with abundant food. A large male panther will kill not very heavy cattle; but as a rule they confine themselves to the smaller animals mentioned. They seldom reside very far from villages, prowling around them at night in search of prey, and retreating to their fastnesses before daybreak. Unlike the tiger, they care little for the neighborhood of water, even in the hot weather, drinking only at night, and generally at a distance from their midday retreat."

The scourge that a man-eating panther becomes, and those traits which make him worse than either the lion or tiger when he has taken to preying upon human beings, have been already given at some length; the following statements, however, also by Forsyth, place the panther's enterprise and hardihood before us very vividly: —

"In my early hunting days I fell into the mistake of most sportsmen in supposing that the panther might be hunted on foot with less caution than the tiger. On two or three occasions I nearly paid dearly for the error, and I now believe that the panther is really by far a more dangerous animal to attack. He is, in the first place, much more courageous. For, though he will generally sneak away unobserved as long as he can, if once brought to close quarters he rarely fails to charge with the utmost ferocity, fighting to the very last. He is also much more active than the tiger, making immense springs clear off the ground, which the other seldom does. He can conceal himself in the most wonderful way, his spotted hide blending with the ground, and his lithe, loose form being compressible into an inconceivably small space. Further, he is so much less in depth and stoutness than a tiger, and moves so much quicker, that he is far more difficult to hit in a vital place. He can also climb trees, which the tiger cannot do, except for a small distance up a thick, sloping trunk. A few years ago a panther thus took a sportsman out of a high perch on a tree in the Chindwárá district. And, lastly, his powers of offence are scarcely inferior to those of the tiger himself, and are amply sufficient to be the death of any man he gets hold of. When stationed at Damoh, near Jubbulpúr, with a detachment of my regiment, I shot seven panthers and leopards in less than a month, within a few miles of the station, chiefly by driving them out with beaters; all of them charged that had the power to do so, but the little cherub who watches over 'griffins' got us out of it without damage either to myself or the beaters. One of the smaller species [Forsyth means a leopard, which, together with Byth, Jerdon, and other naturalists, he regarded as a true panther of less dimensions than the other variety], really not more than five feet long, I believe, charged me three several times up a bank to the very muzzle of my rifles (of which I luckily had a couple), falling back each time to the shot, but not thinking of trying to escape, and died at last at my feet, with her teeth fixed in the root of a small tree.

"Another jumped on my horse, while passing through some long grass, before it was fired at at all; and after being kicked off, charged my groom and gun-carrier, who barely escaped by fleeing for their lives, leaving my only gun in possession of the leopard. I had to ride to cantonments to get another rifle, and gather together some beaters. When we returned I took up my post on a rock that overlooked the patch of grass, and the beaters had scarcely commenced their noise when the leopard went at them like an arrow. An accident would certainly have happened this time had my shots failed to stop this devil incarnate before she reached them. She had cubs in the grass, which accounted for her fury; but a tigress would have abandoned them to their fate in a similar case. The last I killed was a man-eater, that took up his post among the high crops surrounding a village, and killed and dragged in women and children who ventured out of the place. He was a panther of the largest size, and had been wounded by a shikari from a tree, . . . rendering him incapable of killing game. I was a week hunting him, as he was very careful not to show himself when pursued, and at last I shot him in a cow-house into which he had ventured, and killed several head of cattle before the people had courage to shut the door."

Among other peculiarities, says Forsyth, "their indifference to water makes it extremely difficult to bring them to book; and indeed panthers are far more generally met with by accident than secured by regular hunting. When beating with elephants they are very rarely found, considering their numbers; but they must be very frequently passed at a short distance unobserved, in this kind of hunting. In 1862, I was looking for a tigress and cubs near Khápá on the Lawá River in Bétúl. Their tracks of a few days old led into a deep fissure in the rocky banks of the river, above which I went, leaving the elephant below, and threw in stones from the edge. Some way up I saw a large panther steal out at the head and sneak across the plain. He was out of shot, and I followed on his tracks, which were clear enough for a few hundred yards, till, at the crossing of a small rocky nálá (gulch) they disappeared. I could not make it out, and was returning to the elephant, when I saw the driver making signals. He had followed me up above, and had seen the panther break back along the little nálá which led into the top of the ravine, and re-enter the latter. I then went and placed myself so as to command the top of this ravine, and sent people below to fling in stones; and presently the panther broke again in the same place, this time galloping away openly across the plain. I missed with both barrels of my rifle, but turned him over with a lucky shot from a smooth-bore at more than two hundred yards. I then went up to him on the elephant, and he made feeble attempts to rise and come at me, but he was too far gone to succeed.

"The panther will charge an elephant with the greatest ferocity. In 1863, near Jubbulpúr, a party of us were beating a bamboo cover for pigs, with a view to the sticking thereof (that is to say, riding them down and spearing them from the saddle); my elephant was with the beaters, when a shout from the latter announced that they had stumbled on a panther. They took to trees, and I got on the elephant to turn him out, while the rest exchanged their hog spears for rifles and surrounded the place. She got up before me, bounding away over the low bamboos, and I struck her on the rump with a light breech-loading gun as she disappeared. Several shots from the trees failed to stop her, and she took refuge in a very dense, thorny cover on the banks of a little stream. Twice I passed up and down without seeing the brute, but fired once into a log of wood in mistake for her, and was going along the top of the cover for the third time, when the elephant pointed down the bank with her extended trunk. We threw some stones in, but nothing moved, and at last a peon came up with a huge stone on his head, which he heaved down the bank. Next moment a yellow streak shot from the bushes, and levelling the adventurous peon, like a flash of lightning came at my elephant's head, but just at the last spring, I broke her back with the breech-loader, and she fell under the elephant's trunk, tearing at the earth and stones and her own body in her bloody rage.

"The method usually resorted to by old Bamanjee and other native shikáris for killing panthers and leopards, was by tying out a kid, with a line attached to a fish-hook through its ear, a pull at which makes the poor little brute continue to squeak, after it has cried itself to silence about its mother. No sentiment of humanity interferes with the devices of the mild Hindu. A dog in a pit with a basket work cover over it, and similarly attached to a line, is equally effective. I have known panthers repeatedly to take animals they have killed up into trees to devour, and once found the body of a child that had been killed by a panther in the Bétúl District, so disposed of in the fork of a tree. They are very often lost, I believe, by taking unobserved to trees. Beating them out of cover with a strong body of beaters and fireworks is, on the whole, the most successful way of hunting these cunning brutes; but it is accompanied with a good deal of risk to the beaters, as well as to the sportsman if he is over-venturesome; and it is liable to end in disappointment in most instances. My own experience is that the majority of panthers one finds, are come across more by luck than good management.

"A large panther was making himself very troublesome . . . in the neighborhood of the Jubbulpúr and Mandlá road. He had killed several children in different villages, and promised, unless suppressed, to become a regular man-eater. I encamped for some days in the neighborhood of his haunts, and the very first night the villain had the impudence to kill and drag away a good-sized baggage pony out of my camp. The night being warm I was sleeping outside for the sake of coolness, and was awakened by a riving and gurgling noise close to my bed. It was too dark to see; so I pulled out the revolver, that in those uncertain times always lay under my pillow, and fired off a couple of shots to scare the intruder. Getting a light, I was relieved to find it was only the pony." This animal did not return to its "kill," and Captain Forsyth's watch was in vain.

There are certain writers, William H. Drummond, and Sir William Cornwallis Harris, for example, from whose works it might be inferred that in East Africa panthers and leopards were of a quite different character from their Asiatic allies. Taking the evidence on record with regard to this continent as a whole, the discrepancy disappears, however, and *Felis pardus* there, appears in much the same aspect as elsewhere. The animals are necessarily modified to some extent by differences implied in a change of province, but in the main they are reported by observers as exhibiting like traits, and performing much the same exploits with those that have been given.

THE JAGUAR

FELIS ONCA, generally called the jaguar, and very often, in the regions he inhabits, *el tigre*, or the tiger, is a large and heavy animal; coming, in respect to its average size, between the Asiatic panther and lion. It is, perhaps, the most exclusively inter-tropical form among *Felidæ*, — or at least the larger species of that family; and although it passes beyond equatorial latitudes both to the north and south, and is found at considerable elevations where the temperature is low, this beast is essentially an inhabitant of hot countries.

H. H. Smith and others look upon the black jaguar of the Brazilian highlands as a distinct species, and one whose range is different from that of the spotted animals of the Amazon valleys and basin of La Plata. W. N. Lockington ("Standard Natural History") is one of several authorities who consider that there may be several true species of *Felis onca*, besides geographical varieties. In short, the zoölogy of this great American cat is not settled, and the records relating to its character and habits are rather scanty.

Looking at a full-grown jaguar carelessly, one might mistake it for a large and thick-set panther, with a rather short, clumsy tail, and very massive limbs. But besides that the angular ocelli on its coat — irregular black borders with an enclosed spot of the same color — are not rosettes, the *ensemble* is scarcely the same with that of a panther, although anatomically these species are nearly identical.

The true home of the jaguar is in the great woodlands of the Amazon. "Here," says Lockington, "he reigns supreme; the terror of the forest, as the lion is of the desert, and the tiger of the jungle; the acknowledged and dreaded lord of man and beast." Charles Darwin found this species in the basin of the La Plata River, living in reed belts and around lake shores. Unlike the panther, jaguars cannot live without a constant supply of water. Falconer asserts that in some places these animals subsist chiefly upon fish. At all events, they are very expert in catching them, and fish even in rivers whose banks abound with game.

As a rule, however, that large rodent, the capybara, now the only living representative of an ancient family otherwise extinct, is the American tiger's chief article of food, and Darwin reports a saying among the Indians to the effect that man has little to fear from "el tigre's" attacks where these are plentiful. Another point of resemblance between this beast and the panther is their mutual fondness for monkeys.

Natives believe that the jaguar fascinates them. All instances which have been given of the exercise of this power seem, however, to be susceptible of a different interpretation, and naturalists generally discredit the idea that such an influence is ever exerted. Hypnotic phenomena, however, are actual facts, and it is undoubtedly premature to limit the possibility of their induction to human beings.

Apart from this matter, concerning which there is no certainty, it is a fact that the brutes in question take their prey mostly on the surface of the ground, to some extent in water, and likewise among the limbs of trees. They are indiscriminate feeders, and besides all species of land animals that inhabit their range, both wild and domesticated, they destroy vast numbers of turtles and their eggs, lizards, fish, shell-covered species, and even insects. So long as anything has blood, whether red or white, in its body, it does not come amiss to what Wood calls "the jaguar's ravenous appetite." This trait makes him very destructive, and in some places domestic animals have been extirpated.

The jaguar, although he principally subsists upon game, hunts men also, as might be anticipated both from his size, strength, and family traits. An almost unarmed Indian of these regions is no match for a brute like this, even when provided with the blow-gun used in those latitudes.

Being as lazy as a lion, and from his usually abundant supplies, generally in good condition, the jaguar most commonly ambushes prey. Not always, however, for T. P. Bigg-Wither reports that they have been known to follow upon the trail of companies for days, while awaiting a favorable opportunity to seize one of the party. When "el tigre" designs to make a meal of peccary, the character of that creature compels him to surprise it. This is a very bold and inveterately revengeful animal, and moreover is rarely found except in herds. An attack upon one member of the band is instantly and fiercely resented by all, so that strategy upon the jaguar's part is essential to success.

It is not at all unusual to find people congratulating themselves upon the assumed fact that formidable brutes are unacquainted with their own strength and skill. This is one of the many mistakes made concerning lower animals.

Returning to the jaguar's general description, one of his most eccentric propensities is the pursuit of alligators. The jaguar kills and eats these reptiles from choice; or in many instances, simply bites their tails off and lets them go. H. W. Bates found a recently-killed alligator partly eaten. Orton refers to this habit as well known, and both Smith and Wallace speak of it as a matter of common notoriety.

Like all species among the *Felidæ*, this one is nocturnal. Their "dull, deadly-looking eyes," as Barton Premium describes them, are not adapted to excess of light. In remote and secluded places, however, and in the dark recesses of a tropical forest they prowl at all hours, and the author has met with these beasts in the full glare of a vertical sun.

When a jaguar sets out on a foraging expedition at night, he begins to roar like the lion as he leaves his lair; and again like his majesty, he keeps this up at more or less regular intervals until he actually begins to hunt. Jaguars are noisy animals at all times, says Darwin, but they are especially so upon stormy nights, when their "deep, grating roar" reverberates through the forest in a manner very impressive to those unaccustomed to the sound.

Like all animals with retractile claws, they are in the habit of sharpening them, as it is called; but it is not for the purpose of putting a point upon his talons that a jaguar draws them through the bark of trees. All the cats are given to trying how far they can reach, and all of them, both in killing game and feeding, get their nails clogged with shreds of flesh. It is to cleanse them that they scratch tree trunks, from time to time, as they go along. Darwin asserts that each animal has an especial tree to which he resorts for this purpose.

It is agreed among several authorities that a jaguar constantly strikes down, disables, and kills game with a blow of his massive forearm. At the same time, Wood, Humboldt, and Holder write as if death always ensued from dislocation of the neck. When a horse or some other large quadruped is seized, says the former, his assailant "leaps from an elevated spot upon the shoulders . . . places one paw on the back of the head and another on the muzzle, and then with a single tremendous wrench breaks the neck." So far as the act described is assumed to be of invariable occurrence many equally reliable accounts differ entirely, and the author knows from personal experience that jaguars will attack in front, make their assault on level ground, and in some instances do not attempt to kill either man or beast by forcing back the head.

Independently of other facts and considerations which bear upon this brute in its relation to man, the name by which it is known among the natives is more conclusive with regard to character than a host of witnesses. According to Burton the word *jaguar* is composed of the Indian (*Tupi*) *ja*, we or us, and *guara*, an eater or devourer; and it may be assumed that when tribes of savages conferred such a designation as this, they had very good reason for doing so. It may be said, however, that other etymologies of the word have been given.

In the olden days of exploration, both Gonzalo Pizarro and Orellana spoke of the loss of human life from the depredations of jaguars; but, strange to relate, their successors, the accomplished missionary priests, Artiega and Acuna, have nothing to say about them in their sketch of the natural history of Northern Brazil.

Like tigers, lions, and panthers, the jaguar, no doubt, finds it easier to kill a man than almost any other animal that will afford him a full meal, and under favorable conditions he acts accordingly. Hence along the Brazilian frontier of Guiana where these beasts are very numerous, E. F. im Thurn relates that he found the forest tribes sleeping in hammocks swung high enough above the ground to be out of reach of a spring. J. W. Wells and the distinguished Waterton describe the way in which their tents were beset by jaguars. Humboldt tells how his mastiff was carried off from within his camp on the Rio Negro. Darwin mentions that "many woodcutters are killed by them on the Paraná," and that they "have even entered vessels at night," and Von Tschudi recounts how one broke into an Englishman's hut, seized his boy, and bore him off into the forest.

When we examine the records of the first European travellers in those provinces infested by jaguars, their testimony with regard to its character is quite unanimous.

In the Adelantado Pascual de Andagoya's narrative of Pedrarias Davila's expedition he says, "there are lions and tigers" — by which all the Spanish and Portuguese writers meant pumas and jaguars — "on the Isthmus of Panama, that do much harm to the people, so that on their account the houses are built very close to one another, and are secured at night." Father José de Acosta ("Historia natural y moral de las Indias") explains, however, that these beasts are not equally dangerous. "The tigers are fiercer and more cruel than the lions." Likewise it is more perilous to come in their way "because they leap forth and assail men treasonably."

Pedro Cieza de Leon, of whom Prescott remarks that "his testimony is always good," gives an account of the state of affairs on the road between Cali and the port of Buenaventura. Here are "many great tigers, that kill numbers of Indians and Spaniards as they go to and fro every day." Likewise in the mountainous portions of the district, these animals were so destructive that the Indian houses, which are "rather small, and roofed with palm leaves, . . . are surrounded by stout and very long palisades, so as to form a wall; and this is put up as a defence against the tigers." So far as the author's acquaintance with the Spanish and Portuguese relations goes, all authorities of this class agree in giving these beasts the traits that those theoretical and practical considerations mentioned respecting the temper and habits of the large carnivora would lead us to look for.

The writer never saw a full-grown animal of this kind which had been domesticated to the extent of being harmless if left at large, and never succeeded in taming one completely himself. E. George Squier ("Adventures on the Mosquito Shore") mentions an incident in which such was the case. He was summoned to an interview with "The Mother of the Tigers," who, under this ominous title, proved to be a modest young Indian girl, and the high priestess of one of those secret semi-religious societies that gave Alvarado so much trouble in the days of the Spanish invasion. Her retreat lay in the darkest recesses of one of those gloomy forests where there is always twilight, even at the tropical noonday. He found that Sukia was only attended by one old woman, and guarded by an immense jaguar. This beast did not like the stranger's appearance, but made no attack, and at once passed into the house and lay down when commanded to do so.

Perhaps it is unnecessary to bring, as might readily be done, proof of what might be assumed beforehand; namely, that an animal like the jaguar is certain to attack men wherever their possession of firearms has not in the course of time taught the sagacious beast that the contest is an unequal one. It happens, however, that the explorer C. Barrington Brown ("Canoe and Camp Life in British Guiana") has given some quite explicit information concerning a point which has been rarely touched upon, that is to say the behavior of a wild beast that very probably never saw a man before, and certainly not a white man.

Brown was in a country infested by jaguars, but while remaining in the peopled regions he does not say much about them. Once, however, he records the fact that he encountered an Indian whose neck was much distorted by a bite received from this animal. The man was accompanied by a friend armed with a gun when the jaguar sprang upon him, and was shot dead by his friend. Most of Brown's explorations were made in boats, by the waterways of the Essequibo, Corentyne, and other rivers and their affluents. He penetrated into parts which were, so far as human beings are concerned, nearly or entirely uninhabited.

"On one occasion," says this author, "when we had landed and were pursuing a herd of bush-hogs," — peccaries, he means, — "two men were left in charge of the boat. We had not been away in the forest more than two or three minutes, when these men heard a heavy footfall on the bank above them, and looking up saw a large jaguar gazing down upon them from the very spot up which we had clambered." In other words, neither the sense of smell, nor actual sight, taught him anything about those enemies whom he, in common with all other wild beasts, is so generally represented to fear instinctively. "They immediately pushed the boat off shore, fearing an attack from the tiger." Afterwards his men told Brown "that this animal was one of those the Indians call 'Masters of the herd,' that it followed herds of swine wherever they went; and that whenever it was hungry, and found a pig at a little distance from the rest, pounced upon it, killing it with one blow of its huge paw. The squeak of the stricken one always brought down its companions to the spot, whereupon the jaguar climbed a tree for safety till the storm it had brewed was over, and the pigs left the spot; then it descended from its perch to feed on the flesh of its victim. . . .

"In ascending that portion of the Corentyne below Tehmeri rocks, we saw a large jaguar standing on a granite rock close to the river bank, which immediately bolted into the forest as we paddled to the spot. Glancing up at the place where it had disappeared, I saw it sitting down and gazing intently at us, without showing the least sign of fear. I took aim behind the shoulder and fired a charge of large shot, which caused it to bound forward, fall and roll over. But at once regaining its feet it made off into the forest." Although they followed the bloody trail, the animal was not seen again.

Brown had four other shots at jaguars — all of them close — and he wounded two, but never succeeded in bagging a single one. In every case observed by him there was an entire absence of that behavior which is said to be natural and instinctive. The animals he saw expressed only wonder at the sight and scent of man, as well as at the sound of his voice.

Father Acosta declares that the jaguar attacks "treasonably," that is to say, being treacherous like all cats, and one of the laziest of animals besides, he springs upon his prey, as a rule, from an ambush, which may be above the creature seized or on a level with it, according to circumstances.

Like all large beasts of prey, these brutes kill in a variety of ways as existing conditions and the size and structure of the creature assaulted suggest, — they break its neck, tear open the blood-vessels in its throat, strike it dead with a blow from their powerful and massive forearms, crush its life out in their spring, drown it, and tear it to pieces while alive. This last is the way in which such vast numbers of the great river turtles are destroyed: they are turned upon their backs, the claws inserted beneath the breast plate, and these unfortunates are then torn asunder.

With reference to the act of overwhelming an animal, crushing it to death, or killing it by shock, Emmanuel Liais ("Climats, Géologie, Faune, du Brésil"), who gives a somewhat different etymology for the word *jaguar* from that before mentioned, remarks that this term may be translated in a way that refers directly to its method of taking life. "*Le nom de Jaguâra peut alors se traduire en français par le périphrase: Carnassier qui écrase sa proie d'un seule bond.*" This plan is, however, inapplicable to large game.

When a jaguar catches fish, either by waiting till they rise, or by attracting fruit-eating species by tapping with his tail so they think food is falling from the trees, he simply tosses them on shore, and they suffocate in the air; but with the lemantin of the Amazon, upon which he constantly preys, that would be impossible. Paul Marcoy saw the act of capture and describes it in these terms: "At the distance of twenty paces, on a bank facing us, and but a few feet in height, a jaguar of the larger species, — *Yahuaraté pacoa sororoca*, — with pale red fur, and its body beautifully marked, was crouching with fierce aspect, on its fore-paws, its ears straight, its body immovable. . . . The animal's eyes, like two disks of pure gold, followed with inexorable greed the motions of a poor lemantin which was occupied in crunching the stalks of false maize and water-plantains that grew on the spot. Suddenly, as the lemantin raised its ill-shaped head above the water, the jaguar sprang on it, and burying the claws of his left paw in the neck, weighed down the muzzle with those of the right, and held it under water to prevent its breathing. The lemantin, finding itself nearly choked, made a desperate effort to break loose from its adversary, but he had no baby to deal with. The tiger was now pulled under and now lifted out of the water, according to the direction of the violent somersaults of his victim, yet still retained his deadly hold. This unequal struggle lasted some minutes, and then the convulsive movements of the lemantin began to relax, and finally ceased altogether — the poor creature was dead. Then the jaguar left the water backwards, and resting on his hind quarters, with one fore-paw for a prop, he succeeded in dragging the enormous animal up the bank with the other paw. The muzzle and neck of the lemantin were torn with gaping wounds. Our attention was so fixed and close — I say *our* advisedly, for my men admitted that they had never seen a similar spectacle — that the jaguar, which had just given a peculiar cry, as if calling his mate or his cubs, would shortly have disappeared with his capture, had not one of the rowers broken the charm by bending his bow and sending an arrow after the cat, which, however, missed its mark and planted itself in a neighboring tree. Surprised at this aggression, the animal bounded on one side, and cast a savage glance from his round eyes — which from yellow had now become red — at the curtain of willows that concealed us. Another arrow, which also missed its object, the shouts of the oarsmen, and the epithet '*sua — sua*,' double thief, which Julio cried at the top of his voice, at length caused it to move away."

It is not from the jungle only, or the fringing reeds of streams, from dense woodlands, or the undergrowth and high grasses of those *restingas* (open spaces amid overgrown and often submerged country), where Bates says they may be most successfully hunted with beaters, that the jaguar bounds upon his prey. He is by no means exclusively a denizen of the forest, and Romain d'Aurignac ("Três Ans chez les Argentins") merely expresses a commonly known fact when, speaking of the pampas, he remarks that "*les jaguars . . . abondent également dans ces parages.*" On these great plains the jaguar subsists upon cattle, horses, and mules, that are to be found there in immense numbers, as well as upon those wild animals whose habits of life confine them to open places.

C. B. Brown, speaking of the causes, whatever these may be, which prevent the increase of jaguars, remarks that "they have no enemies." This is true in so far as there is no single creature except man in those provinces through which they range that willingly comes into collision with them. No doubt the jaguar frequently meets with a violent death, however, which is not inflicted by human agency. In one case that is certain; the great anteater, or ant-bear, has been known to kill him. Wallace and others vouch for the truth of this, and there is nothing intrinsically improbable in the statement that an animal so large, so powerful, and so formidably armed with claws which are more effective than those of the jaguar in every way, might be able to cling to its enemy long enough to inflict mortal wounds. When attacked by a tiger, the ant-bear turns upon his back and uses his talons with deadly effect. It is said that both parties in such an engagement are apt to perish. The jaguar cannot disengage himself, and the ant-eater dies under the fangs of his adversary.

Those qualities which this creature exhibits in procuring food — the varied styles of attack and modes of destruction it makes use of — entitle the American tiger to be considered as among the first of the whole group of beasts of prey. But there is little doubt that some things are attributed to him through that admiration and reverence he excites in the aborigines, which are without foundation. It is said, for instance, that jaguars mimic the cries of many animals, and thus beguile them within their reach. Of those creatures upon which jaguars prey most constantly, however, a number only call at certain seasons, others are practically voiceless, and some, as monkeys in general, are not to be deluded in this manner.

Priests, naturalists, and geographers, whose special pursuits occupied them fully, have chiefly written of the jaguar's provinces; so that the strong light which is cast upon the character and habits of wild beasts by narratives of the chase, is almost entirely wanting. J. W. Wells ("Three Thousand Miles through Brazil") says, speaking of hunting jaguars with dogs, what the author knows to be true;

namely, that animals employed in this way, and in fact the whole canine family in those latitudes where these animals are found, stand in mortal fear of them. He admits, however, that the ordinary Indian dog will not keep upon a tiger's trail without constant encouragement, and that they never close with them. After having been barked at, one can hardly say chased, for a certain distance, this lazy, short-winded brute gets into some large tree and tries to conceal himself, while the curs yelp around it until their noise brings the huntsmen to the spot. That is the *theory* of this proceeding, but practically it does not work, and few jaguars are killed in this manner. Following up a tiger with dogs just in front — for they will not, as a rule, keep upon the trail by themselves — does well enough to talk about; but when the place where this is to be done is a tropical forest, it will be found impossible to put in practice. If the beast were not disposed to come to bay, it might easily get through mazes impenetrable to men, and go its way along paths by which its pursuers could not follow. There is a breed called "tiger dogs" in Mexico and Central America, but the author has never seen them at work, and also knows that the *tigreros*, or professional tiger-hunters of those parts, kill most of their game without such aid.

Jaguars are constantly seen abroad by day in remote regions; but from the reports of native hunters, and on the ground of personal observation, the author is inclined to believe that their roar is seldom heard except at night. Waterton speaks of it as an "awfully fine" sound, and says that "it echoed among the hills like distant thunder."

Some travellers describe it as a deep, hoarse, rapid repetition of the syllables *pa-pa;* and Brown, referring to the calls of two jaguars he heard on the Berbice River, thought their "low, deep tones," which "made the air quiver and vibrate, . . . had a grand sound, with a true, noble ring in it." The writer never detected anything like a "ring" in it; on the contrary, the ordinary intonation is markedly flat, like that of the panther's and tiger's ordinary cry. A jaguar can roar, however, and often does so with violence: under all modulations his tones convey the impression of great power.

The question how far jaguars hunt by scent, and how far by sight, could not probably be answered, both senses being constantly employed. T. P. Bigg-Withers relates that one of them trailed him "all day waiting for a favorable opportunity" to attack, and that a Botocudo Indian was finally seized, but escaped with some comparatively trifling injuries. This pursuit was carried on no doubt chiefly by scent, although the animal had been seen more than once. Major Leveson ("Sport in Many Lands") makes a statement in connection with shooting from machans to the effect that elevated positions are favorable to the sportsman because wild beasts "never look up." He excepts leopards, it is true, but the fact is that all *Felidæ*, leaving out lions and tigers, which are too heavy and large to climb, use their eyes in every direction, and in prowling for food through forests, scrutinize the trees where their prey is often found, as closely as they do surrounding jungle and open spaces. Those natives who live among tigers on this continent do not at all events think themselves safe in trees, since E. F. im Thurn and others explain that they not only swing their hammocks out of reach among branches, but build fires around the stems to prevent them from being ascended. In such a case the jaguar would probably act as he does when a monkey gets out to the end of an isolated limb that will not bear his weight — that is to say, spring upon the prey, and come to the ground with it.

When a lion or tiger receives a shot, it is very often replied to by a roar, and this whether the animal attacks in return or bounds away. A jaguar, however, generally bears his wounds without any outcry, and if he intends to fight, does so, like the panther, at once. The writer has neither seen nor heard that these animals make use of those stratagems that tigers constantly, and lions frequently, adopt for the purpose of intimidating their assailants and causing them to retreat. It would appear that jaguars do not commonly make feigned assaults, but generally charge in earnest, with lightning-like rapidity, and desperate determination. The writer, speaking from experience, is inclined to think that these animals act in this way as constantly as the panther. There may be, however, numerous exceptions to this behavior; the opinion expressed is not offered as if it were final, and the data upon which it is based are extremely imperfect. More than that, it should be acknowledged with regard to any facts stated, that they only represent this, or any other animal's average behavior. There can be no doubt that wild beasts will sometimes do anything and everything which is not positively impossible.

Whether the current opinion that black jaguars are more ferocious than those of the spotted variety be true, the author is not able to say. Among *tigreros* this is believed to be the case; but that kind of animal is rarer than the others, attracts more attention, and being undoubtedly dangerous, naturally gathers round it certain superstitions with which the minds of this class of men become impregnated. Natives, in general, do not appear to make any particular distinction between the varieties, and such records as we possess place them very much upon a par, with regard to the habits and characteristics that have been spoken of.

The jaguar's strength is very great. These beasts are well known to "carry off," as it is called, the bodies of horses, etc., that have been killed. They swim broad rivers also, and are said, like the royal tiger, to fight effectively while in the water. Wood quotes Dr. Holder to the effect that on one occasion a jaguar destroyed a horse, dragged it to the bank of a large stream, swam across with his prey, and finally conveyed it into the forest. The writer in the "Encyclopædia Britannica" refers to the same story, but besides these authorities, this kind of an exploit has not been recorded by any one.

Darwin states that the jaguar prowling at night is much annoyed by foxes, that attend his movements and keep up a constant barking. It is well known that jackals follow or accompany lions under like circumstances, and Darwin speaks of this parallel association as a "curious coincidence." But the fox is in this case an interloper like the other, an unwelcome hanger-on in expectation of offal, that betrays the jaguar's presence when he, usually a noisy animal, has cause to be quiet.

It is singular that a creature so noteworthy, and one so frequently mentioned, should remain so imperfectly known in many important particulars relating to its natural history, habits, and character. Dr. Carpenter ("Zoölogy") remarks that it "may be regarded as the panther of America," and many traits which favor this likeness have been given. It remains to say, however, that while zoölogists express themselves in guarded terms with respect to species of *Felis onca*, and the natives discriminate half a dozen among the spotted kind alone; while Liais describes "*le jaguar noir*" as "a third species," and Azara ("Descripcion y Historia del Paraguay") writes of a yellowish-white variety as a fourth specific form, the black jaguar, in all probability, only adds another to the many resemblances that liken this beast to the panther. Black or dark-brown cubs have not, as in the case of *Felis pardus*, been found, so far as the writer knows, in one litter with those marked with spots; but there is reason to believe that they occur in this manner.

Two cubs are born together as a rule, although, as happens with other species of this family, the average number is sometimes exceeded. Of the young jaguar's first essays in life very little is known. Whether its father takes part in the whelp's education, as a lion does, or is on the contrary a destroyer of his male offspring, like the tiger; how long parental care continues, and in fact all details relating to its period of infancy, remain obscure. If one inquires about these matters from natives, they entertain him with romances, legends, and folk-lore tales. It was a subject for comment among the early Spanish writers that so few of these animals were killed by Indians. In his "Brief Narrative of the Most Remarkable Things that Samuel Champlain observed in the Western Indies," we find a mention of some jaguar skins that had been bartered by natives, referred to as rarities. Now, as many or more come annually from Buenos Ayres alone as were once procured in the same time throughout the Amazon valleys. Notices of jaguars being taken in traps are occasionally found in books, but detailed descriptions of the process of catching them the author has not met with. Some of the tribes possess efficient weapons of their kind—bows, strong enough, as Cieza de Leon asserts, "to send an arrow through a horse, or the knight who rides it." These Indians are in the habit likewise of poisoning their arrow-heads. Cieza gives an account of how, after much trouble and persuasion, he induced the aborigines at Carthagena and Santa Martha to show him their mode of preparing poison. His relation, however, is not very instructive. Humboldt and Bonpland ("Voyage, etc., Relation Historique") give "*curare*" as the active principle of those mixtures made by Amazonian tribes. These poisons contain, both in South America and all over the world where they are used, matters which are more or less inert, and have been introduced upon purely magical principles. E. F. im Thurn found the effective constituent used in Guiana to be "Strychnos-Urari, Yakki, or Arimaru—*i.e., S. toxifera, S. Schomburgkii, S. cogens.*" Both he and Sir R. Schomburgh speak of other ingredients

— bark, roots, peppers, snake venom — compounded with the more active principle. Waterton gives much the same account of the toxic agent used by means of the bow or blow-gun, and of course there is no doubt that a jaguar inoculated with enough *curare* would die.

As for foreigners, their reliance has always been upon firearms, ever since the first arquebuses were introduced into Spanish America by the *conquistadores;* and nothing less efficient is likely to avail against an animal that Audubon and Bachman say "compares in size with the Asiatic tiger," and is his "equal in fierceness."

THE TIGER

A TIGER to the majority of men is probably the most impressive and suggestive of all animals. Apart from those traits so obvious in his appearance that they affect every one, most beholders have in their minds some material with which imagination works under the quickening influence of his deadly eye. No creature matches him in general powers of destruction; none enacts such tragedies as he, amid scenes so replete with a various interest; none sheds so much human blood.

The hunter's spirit natural to our remoter ancestors survives in their descendants, and few persons are placed under circumstances favorable for its revival without experiencing something of its force. When tigers are the objects of pursuit, this often becomes a passion.

One can scarcely look upon the poor, dispirited wretch behind the bars of a cage, without freeing it in fancy, and transferring the animal to fitting surroundings, — open spaces in jungle, where tall jowaree grass waves in the evening air, deep nálás clothed with karinda and tamarisk, vast, gloomy forests of sál and teak, magificent mountain buttresses, upon whose crags stand the ruined fortresses of long-forgotten chiefs. The tiger of the mind, splendid and terrible is there, and we are there to meet him.

"In some parts of India," remarks Inglis ("Work and Sport on the Nepaul Frontier"), "notably in the Deccan, in certain districts on the Bombay side, and even in the Soonderbunds, near Calcutta, sportsmen and shikáris go after tigers on foot. I must confess that this seems to me a mad thing to do. With every advantage of weapon, with the most daring courage, and the most imperturbable coolness, I think a man no fair match for a tiger in his native jungles." The list of killed and wounded shows that this opinion is not without foundation; and when we consider what it means to meet such adversaries as these on level ground, and face to face, our judgment of its accuracy cannot be doubtful. Gérard compared a contest on foot with a lion to a duel between adversaries armed with equally efficient weapons, but one naked and the other covered with armor in which there were only one or two spots that were not impenetrable. He intended to illustrate, not the animal's invulnerability, of course, but the fact that its tenacity of life was such that, unless instantly killed, it would almost certainly kill its opponent. For this reason sportsmen mostly shoot from howdahs, or machans in tree-jungle. In its depths a great forest is nearly lifeless at all times. In India its skirts are commonly fringed with scrub, and there most of the vitality of these regions concentrates itself. The intense heat of noonday at that season when tiger-hunting begins — namely, in April — makes those immense woodlands as silent and lonesome, to all appearance, as if the hand of death had been laid upon them. But when the short twilight of low latitudes deepens into gloom, the air, before vacant, except for the wide sweep of some solitary bird of prey, is filled with the voices of feathered flocks returning to their roosts. Flying foxes cross vistas still open to the view, and great horned owls flit by on muffled wings. Those spectral shapes which haunt such scenes appear amid the solemn gathering of shadows — contrasts in shade indescribably altering objects from what they are, waving boughs and rigid tree trunks that start into strange relief in changing lights, the distorted forms of animals indistinctly seen moving stealthily about. Throughout those provinces where the most famous tiger haunts are found, positions of advantage, each beetling cliff and isolated hill, holds mementos of the past which are now inexpressibly desolate; the former strongholds of Rajpúts that may, like the Baghél clan, have claimed descent from a royal tiger. As we sit aloft watching, a gleam of water, where when gorged the beast will drink, is visible, and towards that also, each with infinite precaution, and guided by senses of whose range and delicacy of perception human beings cannot conceive, the thirsty denizens of this wilderness take their way. When we mark their timid and uncertain steps, and see how often they hesitate and stop and turn aside, the truth that "nature's peace" is only a form of words expressive of our own misconception and blindness reveals itself most impressively. There is no peace. To hunt and be hunted, to slay and be slain, that is the cycle of all actual life.

Here, while the solemn booming of the great rock monkey sounds like a death knell, those tragedies take place which only a hunter beholds. Every creature has its enemy, and there is one abroad in the gloaming from which all fly. Listen! Above the sambur's hoarse bark, the bison's cavernous bellow, and hyæna's unearthly cry, a deep, flat, hollow voice, thrilling with power, floats through the forest. It is a tiger rounding up deer. If he were in ambush, not the slightest sound would betray his presence. Now his roar, sent from different directions, crowds the game together, and puts it at his mercy.

When and in what way will our tiger come? Some of these beasts never return to a "kill," they lap the blood, or eat once, and abandon their quarry altogether. Others consume it wholly in one or several meals, and even after putrefaction has set in. This animal for whom we wait may approach boldly while it is yet light, or wait till darkness falls, and appear at any hour of the night. At its coming it might put in practice every precaution that could be made use of in stealing upon living prey, or walk openly towards the carcass with long, swinging, soft but heavy strides.

Incidents of any special kind, however, reveal the tiger's nature only in part. What sort of a being is this in whole; how much mind does he possess; what are the traits common to his species; and what their individual peculiarities? Do tigers roar like lions and jaguars, and is it probable that their neighborhood would be announced in this manner? Are they in the habit of going about by day; and if not, on what kind of nights is the beast most active and aggressive? How does a tiger take his prey, especially man? How far can one spring; in what way does he kill; what is his mode of devouring creatures? Can tigers climb? How large are they? Will they assail human beings without provocation, or has the aspect of humanity a restraining power over them? May they be met with casually, and at any time? Where are their favorite lairs? Are they brave or cowardly, cunning or stupid, enterprising, adaptive, energetic, or the reverse?

Sanderson declares that the tiger never roars; he grunts according to Major Bevan, and the only approach to roaring Baldwin ever heard, was a hollow, hoarse, moaning cry, made by holding his head close to the ground. Inglis describes the sound as like the fall of earth into some deep cavity, and Colonel Davidson protests that the tiger barks. Pollok, Leveson, Shakespear, and Rice assert that he roars loudly, terribly, magnificently, tremendously; and D'Ewes ("Sporting in Both Hemispheres") states that in comparison with the roar of a tigress he encountered in the jungle between Ballary and Dharwar, "any similar sound he may have heard, either at the zoölogical gardens or elsewhere, was like a penny trumpet beside an ophicleide." All these names are those of men who hold the most conspicuous positions among hunters of large game; all had killed many tigers and often heard the animal's voice.

Much the same contradictory evidence exists with regard to other things. Colonel Pollok assures us that if he trusted to ambushing game to supply himself with food he would starve to death. Captain Rice, a renowned slayer of tigers, lays down the law to this effect, that these brutes never attack except from an ambush.

Without crowding the page with references, suffice it to say that both by day and night, in forests, thickets, and open grass land, tigers have many times been reported by equally reliable witnesses both to stalk their game, and to spring upon it from a place of concealment.

The striped assassin is provided with a jaw and teeth that enable him to crush the large bones of a buffalo. He can strike his claws, as Major Bevan saw him do, through the skull of an ox into its brain, or break a horse's back with a blow of his forearm. How then does he despatch his victims? Their necks are dislocated, says Colonel Pollok; by biting into them and wrenching round the head with his paws, explains Captain Forsyth. Not at all, protests Baldwin;—dislocation is effected by bending the head backward. In neither way, Dr. Jerdon declares;—the animal's neck is always broken by a blow. Sir Samuel Baker adds his testimony to the effect that a tiger never strikes, and Sanderson says "the blow with his paw is a fable." Other authorities maintain that the cervical vertebræ are crushed when the beast, as it always does, bites the back of the neck; and yet others are sure that since he never seizes an animal in this manner, loss of blood is the immediate cause of death, because the great vessels are severed when a tiger, as is his invariable practice, cuts into the throat. Sanderson states that the blood is not sucked, since a tiger could not form the necessary vacuum. In response to this Shakespear and Davidson both *saw* the blood of animals that had been tied up as lures sucked, and Colonel Campbell, Captain Rice, Major Leveson, and others speak of this act as having come under their personal cognizance.

These animals have been so generally credited with great

springing power that the expressions, "tiger's leap," and "tiger's bound," have passed into the colloquial phrases of more than one language. Nevertheless, when the experiences of eye-witnesses of his performances in this way are referred to, nothing but contradictions are to be met with.

Sanderson ("Thirteen Years among the Wild Beasts of India") thinks "the tiger's powers of springing are inconsiderable." Sir Joseph Fayrer ("The Royal Tiger") says that "it is doubtful whether a tiger ever bounds when charging," and Inglis supports him in this particular. Captain Shakespear regarded a machan twelve feet high as perfectly secure, and Captain Baldwin felt that he was safe when fifteen feet above the ground. Moray Brown saw a tiger jump fourteen feet high. J. H. Baldwin ("The Large and Small Game of Bengal") reports a case in which a tiger leaped the stockade of a cattle-pen "with a large full-grown ox in his mouth," and Dr. Fayrer gives, in the work referred to, the only authentic story of a tiger's having taken a man out of a howdah while the elephant was on his feet. Major G. A. R. Dawson describes the accident that occurred to General Morgan from a wounded tigress that sprang across a ravine twenty-five feet wide and struck him down. Captain W. Rice ("Tiger Shooting in India") measured the leap of a tigress he shot, and found it to be "over seven yards."

Professor Blyth and Dr. Jerdon concluded from their researches at the Calcutta Museum and elsewhere that tigers could not climb. It was certainly a very singular conclusion to come to on anatomical grounds; but waiving this point, we have the statements of Inglis and Shakespear to the fact that several were shot in trees. It is not worth while to continue these inquiries as to whether it is possible to discover something certain about tigers from books; on all points connected with them we should find the same discordances.

Although Buffon's extravagances ("Histoire Naturelle") about this brute's disposition need not be seriously considered, — such expressions as "*sa férocité n'est comparable à rien*" meaning nothing, and no creature, for physiological reasons, being capable of remaining in "a perpetual rage," — enough is known about the beast to make it doubtful whether it deserves the "whitewashing" that some have given its character. But if it be granted that tigers possess intelligence, that in many places they have become acquainted with the effects of European firearms, and are not at all likely to mistake an Englishman with a rifle for a Hindu carrying a staff, many things which seem inexplicably at variance will become plain. If rage does not overpower their discretion, they run away when the prospect of certain death stares them in the face. What do they do when it does not? that is the question at present, and the answer is that they act like tigers. This most formidable of beasts of prey is not in the least afraid of a man because he *is* a man; he does not quail at his glance — that enrages him; his voice will not always startle, — it often attracts; nor can the scent of a human being of itself turn him aside — on the contrary, it frequently guides the beast to his prey. So much for the general view; and we may now go into the jungle again and discuss what befalls, in the light of those principles which have been advanced elsewhere. This will be a *duróra* against the tigers of a district, our hunting-grounds lie in historic spots, and the party is accompanied by elephants, baggage animals, attendants, and all the varied appliances that belong to a raid of this kind conducted upon a large scale.

Close to our camp lie the crumbling *eedghas*, shrines, tombs, and fortress palaces of a race of princes now extinct, and seated in a kiosk around whose crumbling walls half-effaced Persian and Arabic inscriptions tell of the beauty of some girl whose bright eyes closed ages ago, and whose career of ineffectual passion finds a fit emblem in the *pishash*, or transient dust column that glides across the plain, let us attempt to forecast the events of to-morrow. More can be foretold than one would suppose. The tiger's size and age, the configuration of the ground, his previous habits of life, and the places where shade and water are to be found, will certainly affect his movements after he has been roused, and when the shikáris come in we shall know all this. Here is the head huntsman now, who comes back from his scout to make a report to the "Captain of the hunt," an experienced sportsman always elected on such occasions to take a general direction of affairs, and manœuvre our elephants in the field. Mohammed Kasim Ali is a typical figure and worth looking at; a small withered being with a dingy turban wound around his straggling elf locks; dressed in a ragged shirt of Mhowa green, and lugging a matchlock as long as himself loaded half way up the barrel. He bears the big bison horn of coarse slow-burning native powder, and a small gazelle-horn primer. His person is bedecked with amulets, and his beard, he being an elderly man, is dyed red — if he were young, it would be stained gray. But despite this man's grotesque appearance, he possesses a profound knowledge of wood-craft, and as a tracker and interpreter of signs, no savage or white prodigy of the wilderness who ever embellished the pages of a certain style of romance can surpass him.

This worthy delivers himself somewhat as follows: "May I be your sacrifice! Whilst searching with eagerness for these sons of the devil, your slave beheld the footprints of a tiger. *Alla ke Qoodrut*, it is the power of God; then why should your servant defile his mouth with lies? These tracks were made by the great-grandfather of all tigers. The livers of Chinneah and Gogooloo turned to water at the sight, but sustained by my Lord's condescension I followed them to a nálá, and he was standing by a pool. Karinda and tamarisk bushes grew more thickly than lotus flowers in Paradise, but I saw clearly that the unsainted beast was bigger than a buffalo bull. His teeth were as iron rakes, his eyes glared like bonfires, and the spirits of those whom he had devoured sat upon his head." This with many aspirations, to the effect that unquenchable fire might consume the souls of the tiger's entire family.

This rhodomontade — quite in keeping, however, with the individual and his country — means that a large tiger was seen, and will be found for us next day.

The one that Kasim Ali, the eloquent, saw by the pool was making ready for his nightly excursion; for although they are frequently seen abroad by day, these animals are nocturnal in habit. The writer, however, sees no reason for repeating a remark which is often made in this connection, namely, that they are "half-blind" during daylight. There is no rigidity in the iris, nothing to prevent the eye from adjusting itself to different degrees of intensity in that medium by which the retina is stimulated. He sees very well at night, and if sensitive to a strong light, so are many other animals whose vision is also good when it is not dark. It is habitual with tigers to seek shade; and any eyes, except those of some birds, would be dazzled by the intense glare of an Indian sun.

When viewed by the shikáris, he had lately roused from his rest as the day declined, and the faint lowing of distant herds, and far-away voices of *Gwallas* bringing home their cattle penetrated to his retreat. He stretched his lithe length and magnificent limbs, his fierce eyes dilated, and a strange and terrible change came over the beast. Every attitude and motion betrayed his purpose. But although murder was in his mind, and all that he did revealed that intention, his movements varied, or would do so, with age and experience. If the animal were young, and had been but recently separated from the tigress, that taught him to find prey, showed how to attack it, and encouraged him to kill for the sake of practice, his actions would exhibit all the boldness that comes from entire self-confidence. He then leaves the lair without precaution, and takes his way through the intricacies of the jungle with confidence, not pausing to examine every sign, as his trail shows. If old, however, an unusual sound would stop him, a footprint in the path that was not there when he last passed would turn him aside. This tiger of ours is not aged, but has learned something since he became solitary like all his kind, except in the brief season of pairing. Experience may be thrown away on men, but not upon tigers. This one will never again make mistakes such as those into which overboldness and want of proper attention have already betrayed him. Once, shortly after he began to shift for himself, a buffalo, of whom he thought that it could be killed as easily as a slim long-necked native cow, tossed him. Another time when too hungry to wait for a favorable opportunity, he seized upon a calf prematurely. No sooner did his roar of triumph as he struck it dead echo through the jungle, than a dark crescentic line fringed with clashing horns confronted him. It came on in quick irregular rushes, and no tiger could withstand such an array, so he had to fly. His glossy hide was ripped likewise by a "grim gray tusker," which the unsophisticated youth designed to despatch without difficulty. Before these instructive incidents occurred something more had been learned also.

One morning the silence was broken by blasts of cholera horns, the beating of tom-toms, and wild cries from a multitude of men — such men, however, as he knew and had frequently observed in the jungle and elsewhere. But there was now a man, mounted on an elephant, the like of which he had never seen, but whose appearance is not forgotten. He had guns far worse than matchlocks, instruments of sudden death that killed his mother. This formidable robber, for all his ferocious temper, great strength, and terrible means of offence, is as cunning as a fox, and wary to a degree that closely simulates cowardice. But one might as well call North American Indians cowards, — which by the way is often done by those whose opinions are unbiassed by any personal acquaintance with them, — because they always fight on the principle of taking the greatest advantage and least risk.

To start a party such as ours takes time, and of the value of time no Hindu has the slightest idea. The mob of beaters are packed off with strenuous injunctions to keep together, but they will not do so. An ineradicable heedlessness besets them, and they are certain to straggle, though the risk that doing so entails is perfectly well understood. The Oriental says, "If it is my fate to perish thus, how can I avoid the decree of heaven? My destiny is fixed; it is in the hands of God, and may the devil take these infidels who talk as if matters could be otherwise than as they are."

Every crupper, breast-band, girth, and howdah cloth must be looked to by the hunters themselves; mahouts and attendants cannot be trusted to equip their charges, and if things were left to them, an elephant would be disabled every day.

All our proceedings as we draw near to the tiger require to be conducted with reference to the lie of the land. Whether he be beaten for with elephants, or roused by the unearthly clamor of the crowd that has come to drive him, it is probable that his first act will be an attempt to escape. He carries a perfect topographical chart of the neighborhood in his head, and an unguarded avenue of egress means that we shall not carry back his spoils. When he does start, it will not be with the wild, affrighted rush of a bison or sambur stag; his retirement, if he is not actually sighted, is made with the deathly silence of an elephant warned of danger. He makes use of every mode of concealment, creeps from bush to bush, from tree to tree, from rock to rock, crouching where cover grows thin or fails, so that the colors of his coat assimilate with those of the herbage, and he becomes well nigh invisible even in places where it seems utterly impossible for so large an animal to hide himself. In denser jungle the fugitive stops and stands with head erect to listen, or rears up amid long jowaree grass, taking in every sight and sound that indicates the position of his enemies. Thus his advance is made towards the point at which it is intended to break away; and if it be necessary to cross bare spots, he does so, not indeed with a panther's lightning-like rapidity, but in long, easy bounds that devour the distance.

Under all circumstances, if the ground is sufficiently broken to permit of it, the tiger keeps among ravines, at one time traversing the crest of a ridge, at another stealing along through the underbrush below. Then it is that the pad-elephants and lookouts in trees come into play in order to turn him in the direction where the rifles are stationed; the former by their presence, the latter by softly striking small sticks together.

It is very likely, however, that the surface may not admit of beating with men; then the sportsmen advance in their howdahs, and one may see how a highly-trained shikar tusker can work.

Sir Samuel Baker ("Wild Beasts and Their Ways") described the qualities of a good hunting animal in action. His party were out near Moorwara. It was in the dry season, and they were keeping on a line parallel with the railroad, and about twenty miles from it. The heat had evaporated tanks, caused upland springs to fail, and dried up pools and watercourses, so that tigers, that cannot endure thirst, were driven from their accustomed retreats into places more accessible. On this occasion the natives were beating towards Baker's elephant, but the beast, as it sometimes does, broke back upon their line at once.

"We were startled," he continues, "by the tremendous roars of this tiger, continued in quick succession within fifty yards of the position I then occupied. I never heard, either before or since, such a volume of sound proceed from a single animal. There was a horrible significance in the grating and angry voice that betokened extreme fury of attack. Not an instant was lost. The mahout was an excellent man, as cool as a cucumber, and never over-excited. He obeyed the order to advance straight towards the spot where the angry roars still continued without intermission.

"Moolah Box was a thoroughly dependable elephant; but although moving forward with a majestic and determined step, it was in vain that I endeavored to hurry the mahout. Both man and beast appeared to understand their business completely, but according to my ideas the pace was woefully slow if assistance was required in danger.

"The ground was slightly rising, and the jungle thick with saplings about twenty feet in height, and as thick as a man's leg; these formed an undergrowth among the larger forest trees.

"Moolah Box crashed with his ponderous weight through the resisting mass, bearing down all obstacles before him as he steadily made his way across the intervening growth. The roars had now ceased. There were no leaves on the trees at this advanced season, and one could see the natives among the branches in all directions, as they perched for safety on the limbs to which they had climbed like monkeys at the terrible sounds of danger. 'Where is the tiger?' I shouted to the first man we could distinguish in his safe retreat only a few yards distant. 'Here! here!' he replied, pointing immediately beneath him. Almost at the same instant, the tiger, which had been lying ready for attack, sprang forward with a loud roar directly for Moolah Box.

"There were so many trees intervening that I could not fire, and the elephant, instead of halting, moved forward, meeting the tiger in his spring. With a swing of his huge head he broke down several tall saplings, that crashed towards the infuriated tiger and checked his onset. Discomfited for a moment, he bounded in retreat, and Moolah Box stood suddenly like a rock, without the slightest movement. This gave me a splendid opportunity, and the .577 bullet rolled him over like a rabbit. Almost at the same instant, having performed a somersault, the tiger disappeared, and fell struggling among the high grass and bushes about fifteen paces distant.

"I now urged Moolah Box carefully forward until I could plainly see the tiger's shoulders, and then a second shot through the exact centre of the blade-bone terminated its existence."

In this attack four men were wounded, but it is not often that a tiger charges home upon a line of beaters; generally, only stragglers suffer, although, as has been said, some tigers attack immediately upon being found. Whenever and however the assault is made, it must needs be a terrible one, and to most creatures at once overwhelming. Imagine a beast like this, so active, so powerful, so armed, — five hundred pounds' weight of incarnated destructive energy launched by such muscles as his against an enemy. "It has been the personification of ferocity and unsparing cruelty," says Sir Samuel Baker. But it is to the terrible character of its attack, to the fact that this is so frequently fatal, and to the awe-inspiring appearance of the beast as it comes on with dilated form and fire-darting eyes, that much of its reputation for more than ordinary ferocity is due. A tiger is beyond question the most formidable of all predatory creatures when earnest in his aggressive intentions; very frequently, however, he is not so. False charges, made in order to intimidate, are more common than real ones. A tiger will bristle, and snarl, and roar, apparently with a perfect consciousness of the additional impressiveness given to his general appearance in this way. Some are, of course, braver than others; locality and their experience of human power make a wide difference between those whose characters have been formed in separate areas. Still everywhere their temper is short and fierce, and when roused ·to fury they fight desperately. When we hear of the abject cowardice of these beasts, — how they slink away from before the face of man and cannot endure his look, how they will never assail him if not provoked, and how they die like curs at last, — it is natural, and a mere suggestion of common sense, to think that these are *ex parte* statements, premature generalizations, sweeping conclusions from special experiences, and misinterpretations of observations that a little diligence and proper intellectual sincerity upon the part of their narrators would have shown to be more than counterbalanced by facts of a different complexion.

No two tigers are identical in anything, and all the elements of uncertainty and dispute which have been specified make their appearance when we come into contact with them. Nobody knows or can know what will happen then. Silently like some grim ghost, the animal may steal within shot, and fall dead at the first fire. Sometimes he bursts from a dense clump of bushes that the hunter's sight has been unable to penetrate, and if hit, rages round the tree from which the ball came as if mad; or, if his foes be within reach, he kills or is killed. Occasionally when not well watched by lookouts, the first intimation that his domain has been invaded is the signal for a retreat to some secure hiding-place, — the pits and passages of an abandoned mine, or a cave perhaps, in which latter case, if it be attempted to dislodge him by an indraught of smoke from fire kindled at its mouth, it will be seen that a tiger can breathe in an atmosphere such as would seem to be necessarily fatal to any animal. Finally, the brute may break back and attack the beaters, or creep through their line, or charge the elephants, and perish amid the wildest display of fury and desperation. Finally, as it sometimes, though rarely happens, the first stir in the jungle sends him off by an unguarded path across ridges and plains to some distant lair, and the hunt for that day is bootless.

Tiger-shooting is never without danger to the sportsman. Many a man has been clawed out of a tree and killed, or caught before he could get out of reach. Elephants have been pulled down, or the howdah ropes have broken and precipitated its occupants into the tiger's jaws. Moreover, nine elephants out of ten are not stanch, they become panic-stricken and bolt; in which event the risk of being dashed to death against a tree is greater than that of any other fatal accident that is likely to occur.

Most accounts of tigers are confined to their connection with mankind; but if this be the more important, it certainly is not the more general relationship. Out of the large number born every year (though not in the same season, for these animals pair irregularly) few come in contact with human beings. They prey upon the larger animals of their respective provinces, both wild and domestic, but, of course, chiefly upon the former. In this way they are of positive benefit to the agricultural class. Baldwin, Sanderson, Leveson and others, whose observations made upon the spot, and with the best opportunities for knowing the truth in this matter, are not likely to be incorrect, state that but for the aid rendered by tigers in keeping down the numbers of grain-eating species, the Indian cultivator would find it almost impossible to live. No doubt the same condition of things prevails in other parts of Asia. Cattle-lifters, however, impose a heavy tax on the country, and as these generally grow fat, lazy, and rarely hunt, they are a decided disadvantage to any neighborhood. Furthermore, it is from among this class that most man-eaters come. In districts to which cattle are driven to graze, and then withdrawn when the grass fails, tigers accustomed to haunt the vicinity of herds, and that have remained for the most part guiltless of human blood so long as their supply of beef lasted, are apt to eat the inhabitants when it fails. One of these marauders upon livestock will kill an ox every five days, and smaller domestic animals proportionately often, and it is easy to see that the cost of supporting them must be very considerable.

So much has been said in connection with other beasts of prey upon the subject of those reports in which each group is represented to have an invariable way of capturing and killing game, that it seems unnecessary to enlarge upon this point with reference to tigers. They stalk animals, and spring upon them from an ambush. When a victim has been caught, it is destroyed by a blow with the arm, its neck vertebræ are crushed by a bite, its throat is cut, or head wrenched round. Very probably the tiger does not strike habitually like a lion. He often does so, however, and the fact that one was seen to drive his claws into the brain of an ox has been mentioned. Sir Joseph Fayrer reports the case of a tiger that dashed into a herd, "and in his spring struck down simultaneously a cow with each fore foot." Major H. A. Leveson ("Hunting Grounds of the Old World") saw one of his men killed in the Annámullay forest in this manner. "His death," says Leveson, "must have been instantaneous, as the tigress with the first blow of her paw crushed his skull, and his brains were scattered about."

"I venture to assert," says Colonel Gordon Cumming ("Wild Men and Wild Beasts"), "that one of the chief characteristics of the tiger is, that in its wild state, it will only feed on prey of its own killing." No other name of equal weight has been appended to a statement such as this. On the contrary, nearly all evidence goes to show that tigers are very indiscriminate in their eating, that they feed on almost anything, living or dead, fresh or putrid. Captain Walter Campbell ("The Old Forest Ranger") mentions the fact of their appropriating game already killed as coming under his personal observation; and Major Leveson ("Sport in Many Lands") records that he shot two tigers in the Wynaad forest while they were engaged in a desperate fight for the possession of a deer's carcass. It is notorious that tigers so constantly destroy their cubs that the tigress leaves her mate almost immediately after they are born, and conceals her young. There are several instances in which she herself has been devoured, and there is no doubt of the cannibalism of this beast. J. Moray Brown ("Shikar Sketches"), speaking of the frequency of combats between tigers, says that, "occasionally the victor eats the vanquished." Colonel Pollok ("Sport in British Burmah") informs us that "when two tigers contend for the right of slaughtering cattle in any particular locality, one is almost sure to be killed, and, perhaps, eaten by the other. I have known instances of this happening." General W. C. Andersson shot a tiger in Kandeish, within whose body he found the recently ingested remains of another, whose head and paws were lying close by in the jungle. General Blake also discovered, near Rungiah in Assam, the partially devoured body of a tiger that had been killed by one of its own kind.

Except incidentally, technical details bearing upon character have not been mentioned; the tiger's size, however, has no doubt a marked influence upon his mental traits. Looking upon a trail that goes straight towards the water, which other creatures approach so differently, one sees how the animal that left those footprints — nearly square in the male, oval in case of a tigress — felt no fear of any adversary, and therefore must have been of considerable bulk. Not only the best authorities, so far as formal zoölogy is concerned, but almost every one who has devoted special attention to this subject, gives the length of an average tiger, when fully developed, at about nine feet six inches from tip to tip. The female is quite twelve inches shorter. Many writers, however, admit the existence of tigers ten feet long, and no one is in a position to deny that some may attain to that length. But when a writer like Sir Joseph Fayrer ("The Royal Tiger of Bengal") says that he has "measured their bodies as they lay dead on the spot where they had fallen," and found them to be "more than eleven feet from the nose to the end of the tail," there is nothing to be replied, except that, very few persons have been so fortunate as to see the like. There was once, indeed, a tiger-slayer who used to shoot specimens fourteen feet long and over, but he died gallantly in battle, and his name need not be given.

With regard to the structure of his brain, the tiger is gyrencephalous; that is to say, the lobes exhibit a certain degree of convolution. It may also be said that the cerebral hemispheres project backwards so as to cover the anterior border of the cerebellum, and that these greater segments of the encephalon are completely connected. The nervous structure is not of the highest type known to exist among inferior animals, but it is quite high enough not to militate against an empirical conclusion that this creature's actions show it to be organically very capable.

Of the details of the every-day life of the tiger we know comparatively little. Thousands of cattle, for instance, are killed every year in India, and yet there is but one narrative, so far as the writer knows, of a tiger having been seen to stalk a quadruped of this kind. It is quoted by J. Moray Brown ("Shikar Sketches") from Captain Pierson's relation of the incident. While hunting in the jungles of Kamptee, he saw from the edge of a ravine on which he was resting, a herd grazing on the ground just below, and a tigress at a little distance reconnoitering. Her choice fell in the first place upon a white cow that was straggling, and she approached till within about eighty yards under cover of the bushes, and then broke into a trot. The cow, however, became aware of her danger, and after standing a moment as if paralyzed with fear, dashed into the midst of her companions. The tigress, which during this time had continued to advance, then charged at once, and "in a few seconds she picked out a fine young cow, upon whose shoulders she sprang, and they both rolled over in a heap. When the two animals were still again, we could distinctly see the cow standing up with her neck embraced by the tigress, which was evidently sucking her jugular. The poor creature then made a few feeble efforts to release herself, which the tigress resented by breaking her neck." Major H. Bevan ("Thirty Years in India") saw a tiger "knock over a bullock with a single blow on the haunch, and seizing the throat, lay across the body sucking the blood." Major Leveson ("Hunting Grounds of the Old World"), while lying out by a pool at night, witnessed the death of a sambur deer that was struck down and instantly killed by a tiger. Various narratives of the tiger's attack might be quoted, but his behavior while stealing upon his prey, the manner in which he seeks for it, and the way in which it is discovered, these are points that we know very little about.

"The tiger is a shy, morose, and unsociable brute," Dr. Fayrer remarks, "but like all animals of high type, the range of individual differences is very great." "Nearly every tiger," observes Moray Brown, "has a certain character for ferocity, wiliness or the reverse — of being a man-eater, cattle-lifter, or game-killer — which is well known to the jungle folk."

The tiger's overlordship of the jungle is not maintained without some reverses. A bear sometimes beats him off, but usually these contests end in the bear's being devoured. Sanderson, together with others, reports this upon personal observation. Wild boars occasionally avenge the death of their fellows. Inglis found the bodies of both combatants lying side by side.

Single buffaloes are killed by a tiger; but when a herd is combined against him, as is always the case when his presence is discovered, he has no chance of success. Inglis ("Work and Sport on the Nepaul Frontier") describes such an event, and as it is the only narrative of this kind the author has met with, his account is given in full.

"One of the most exciting and deeply interesting scenes I ever witnessed in the jungles . . . took place in the month of March, at the village of Ryseree, in Bhaugulpore.

"I was sitting in my tent going over some accounts with the village putwarrie and my gomasta. A posse of villagers were grouped under the grateful shade of a gnarled old mango tree, whose contorted limbs bore witness to many a tufan and tempest which it had weathered. The usual confused clamor of tongues was rising up from this group, and the subject of debate was the eternal '*pice*' [small coins].

"A number of horses were picketed in the shade, and behind the horses, each manacled by weighty chains, with their ponderous trunks and ragged-looking tails swaying to and fro with a never-ceasing motion, stood a line of ten elephants. Their huge leathery ears flapped lazily, and ever and anon one would seize a branch, and belabor his corrugated sides to free himself from the detested and troublesome flies.

"Suddenly there was a hush. Every sound seemed to

stop simultaneously as by prearranged concert. Then three men were seen rushing madly along the elevated ridge surrounding one of the tanks. I recognized one of my peons, and with him there were two cowherds. Their head-dresses were all disarranged, and their parted lips, heaving chests, and eyes blazing with excitement, showed that they were brimful of some unusual message.

"Now there arose such a bustle in the camp as no description could adequately portray. The elephants trumpeted and piped; the syces and grooms came pushing up with eager questions; the villagers bustled about like so many ants roused by the approach of a foe; my pack of terriers yelped in chorus; the pony neighed; the Cabool stallion plunged about; my servants rushed from the shelter of the tent-veranda with disordered dress; the ducks rose in a quacking crowd, and circled round and round the tent; and the cry arose of '*Bagh! Bagh! Khodawund! Arree Bap re Bap! Ram Ram, Seeta Ram!*'

"Breathless with running, the men now tumbled up and hurriedly salaamed; then each with gasps and choking stops, and pell-mell volubility, and amid a running fire of cries, queries, and interjections from the mob, began to unfold their tale. There was an infuriated tigress on the other side of the nullah, or dry watercourse, and she had attacked a herd of buffaloes, and it was believed she had cubs.

"Already Debnarain Singh was getting his own pad-elephant caparisoned, and my bearer was diving under my camp bed for the rifles and cartridges. Knowing the little elephant to be a fast walker, and fairly stanch, I got upon her back, and accompanied by the gomasta and mahout we set out, followed by the peon and herdsmen to show us the way.

"I expected two friends, officers from Calcutta, that very day, and wished not to kill the tigress, but to keep her for our combined shooting next day. We had not proceeded far, when on the other side of the nullah we saw dense clouds of dust rising, and heard a confused rushing, trampling sound, intermingled with the clashing of horns, and the snorting of a herd of angry buffaloes.

"It was the wildest sight I have ever seen in connection with animal life. The buffaloes were drawn together in the form of a crescent; their eyes glared fiercely, and as they advanced in a series of short runs, stamping with their hoofs, and angrily lashing their tails, their horns would come together with a clanging, clattering crash, and they would paw the sand, snort, and toss their heads, and behave in the most extraordinary manner.

"The cause of all this commotion was not far to seek. Directly in front, retreating slowly, with stealthy, crawling, prowling steps, and an occasional short, quick leap or bound to one or the other side, was a magnificent tigress, looking the very impersonification of baffled fury. Ever and anon she crouched down to the earth, tore it up with her claws, lashed her tail from side to side, and with lips retracted, long mustaches quivering with wrath, and hateful eyes scintillating with rage and fury, she seemed to meditate an attack upon the angry buffaloes. The serried array of clashing horns, and the ponderous bulk of the herd appeared, however, to daunt the snarling vixen; at their rush she would bound back a few paces, crouch down, growl, and be forced to move back again, before the short, blundering charge of the crowd.

"All the old cows and calves were in rear of the herd, and it was not a little comical to witness their awkward attitudes. They would stretch their ungainly necks, and shake their heads as if they did not rightly understand what was going on. Finding that if they stopped too long to indulge their curiosity, there was danger of getting separated from the fighting members of the herd, they would make a stupid, lumbering, headlong rush forward, and jostle each other in their blundering panic.

"It was a grand sight. The tigress was the embodiment of lithe savage beauty, but her features expressed the wildest baffled rage. I could have shot the striped vixen over and over again, but I wished to keep her for my friends; and I was thrilled by the excitement of such a novel scene.

"Suddenly our elephant trumpeted, and shied quickly on one side from something lying on the ground. Curling up its trunk it began backing and piping at a prodigious rate.

"'Hallo! what's the matter now?' said I to Debnarain.

"'God only knows,' said he.

"'A young tiger! *Bagh ta butcha*,' screamed our mahout, and regardless of the elephant or our cries, he scuttled down the pad rope like a monkey down a backstay, and clutching a young dead tiger cub, threw it up to Debnarain. It was about the size of a small poodle, and had evidently been trampled by the pursuing herd of buffaloes.

"'There may be others,' said the gomasta, and peering into every bush, we went slowly on. My elephant then showed decided symptoms of dislike and reluctance to approach a particular dense clump of grass.

"A sounding whack on the head, however, made her quicken her steps, and thrusting the long stalks aside, she discovered for us three blinking little cubs, brothers of the defunct, and doubtless part of the same litter. Their eyes were scarcely open, and they lay huddled together like three enormous striped kittens, and spat at us, and bristled their little mustaches much as an angry cat would do. All four were males.

"It was not long before I had them wrapped up carefully in the mahout's blanket. Overjoyed at our good fortune, we left the excited herd still executing their singular war-dance, and the enraged tigress, robbed of her whelps, consuming her soul in baffled fury.

"We heard her roaring through the night close to camp, and on my friends' arrival, we beat her up next morning, and she fell, pierced by three balls, in a fierce and determined charge. We came upon her across the nullah, and her mind was evidently made up to fight."

A tiger may fail in front of a herd, but with stragglers, and there are always such, the case is not the same. He can kill individual buffaloes, or he would not be there, and this is done so quietly and expeditiously that very often the act remains for a time undiscovered. His "fore-paw," observes Inglis, "is a most formidable weapon of attack.

... One blow is generally sufficient to slay the largest bullock or buffalo." Then he reports how a tiger, charging through the skirts of a herd, "broke the backs of two of these animals, . . . giving each a stroke, right and left, as he passed along." Now it is certain that an Asiatic buffalo is quite as large and formidable an animal as the bison; and it may naturally be inferred from this, that most of these latter fare differently from the one Leveson and Burton saw fighting at the Nedeniallah Hills.

Having thus secured a supply of beef, the tiger usually withdraws and waits for night to make his meal. But if he were alone with his victim, if there were no danger of being winded and attacked by its companions, he would act differently, and might eat at once. Inglis does not tell how he became acquainted with the following details, but he states that as soon as his prey is struck down, the tiger "fastens on the throat of the animal he has felled, and invariably tries to tear open the jugular vein." This he does instinctively, because he knows intuitively that "this is the most deadly spot in the whole body." But the tiger's intuitions and Inglis's knowledge are both at fault in this particular. "When he has got hold of his victim by the throat, he lies down, holding on to the bleeding carcass, snarling and growling, and fastening and unfastening his talons." In some instances, continues this writer, he may drink the blood, "but in many cases I know from my own observation that the blood is not drunk." After life is extinct, these brutes "walk round the prostrate carcasses of their victims, growling and spitting like tabby cats." If they wish to eat then, the body is neatly disembowelled, and the meal begins on the haunch. A panther or leopard would commonly commence with the inner part of the thighs, "a wolf tears open the belly and eats the intestines first," and a hawk, and other birds of prey, pick out the eyes; but a tiger follows the course described, as a rule, and after having bolted — for he never chews his food — as much as he can hold, the remainder is dragged off and concealed, or at least this is the intention, though his design is always very imperfectly executed.

Colonel Barras, while waiting for a tiger driven by beaters, saw the beast break back upon their line, as these animals are apt to do, and with evil consequences, seeing that no power can keep Hindus together.

"I saw him rise up on his hind legs and take the head of one of them in his mouth. In an instant he dropped his victim, and made short pounces at the others, who (as may be supposed) were flying wildly in all directions. Numbers of them left the long cloths they wear round their heads sticking to the thorny bushes. These, it seemed to me, the tiger mistook for some snare, as he suddenly turned and bounded away at tremendous speed under the very tree I was in. Owing to the great pace he was going I missed him. I have since seen others miss under the same circumstances, but at the time I felt my position keenly, being under the impression that other persons invariably dropped their tigers whenever and wherever they might get a glimpse of them.

"It only remained now to follow up the brute with elephants. Owing to the fierceness of the sun, he would not be likely to travel far, or make many moves. After tracking for about an hour, he did turn out in front of one of the elephants, and was fired at by the people in the howdah, with what success I do not remember. For a moment he pulled himself up, and seemed about to charge, but thought better of it, and was soon out of sight again. We followed him for some hours along the rocky banks of the river, visiting all the most likely nooks and corners, in hopes that he might find it impossible to travel any further over the burning rocks. Towards evening he was descried at the distance of a quarter of a mile, swimming across a deep pool that led into an extensive piece of forest. Here we deemed it advisable to leave him for the night, and organize a fresh plan for the morrow. Accordingly the next morning a beat was commenced from the opposite side of the wood, which proved successful. The tiger broke readily and was shot by one of the party. It was a very fine male, in the prime of life. At first I wondered why it was so certainly admitted to be the tiger of the day before. On asking the question, his feet were pointed out to me. They were completely raw with his long ramble over the burning rocks. It is not improbable that had he been only slightly driven, he would have travelled miles away during the night, and we might have lost him."

As for the wounded man, whose skull, strange to say, had not been crushed, he was carefully attended to and well rewarded for his sufferings.

"An occasional accident of this sort should not be looked upon as a proof of the brutal indifference of the English in India to the lives of the suffering natives —

quite the contrary. The natives, except under European leadership, will not go out against dangerous animals. Bapoo says, 'My cow is not killed, and besides I have obtained a charm from a holy man, by which she is made safe against tigers. Why should *I* go out?' On the other hand, Luximon says, 'My cow *is* killed; I shall certainly not go.'" In consequence of these reasonings, they and their cattle continue to be eaten. As Barras says, "The result is that the tigers get the better of the natives, and kill so many of them and their cattle, that I have seen many ruined villages, which have been abandoned owing to the neighborhood of these animals. It is, therefore, a very good thing for the inhabitants when a well-appointed shooting party arrives.

"One of the most curious features of tiger-shooting is the extraordinary tenacity with which both the Europeans and natives engaged in the sport adhere to certain traditions. In vain does a tiger break through all established rules before the very eyes of those engaged; the shikáris, both white and black, continue as firm as ever in their articles of faith, and, by their blind belief in the same, often lose a tiger. I propose, therefore, to mention a few of the most cherished laws, and to show in the following pages that they are in every instance fallacies.

"(1) A tiger never charges unless wounded, or in defence of its young cubs.

"(2) It never lies up for the day in hot weather in a jungle where there is no water.

"(3) It never looks upward so as to see any one in a tree.

"I have already given one instance of an unwounded tiger charging and nearly killing a beater, and I now propose to show how another was unprincipled enough to break two of the three rules at the same time.

"A few days after the events narrated in the preceding chapter, I and the four others comprising our party were duly posted across a wide nullah (dry watercourse). Gibbon was told off for a tree growing on the top of the bank. The fork into which he climbed must have been quite twelve feet from the ground, so that as I sat in my bush in the bed of the nullah he appeared almost in another world. As soon as we were all settled the beat began. Our band on this occasion was unusually good. It produced a loud and piercing discord.

"Almost immediately was heard the sound as of a horse galloping down the stony bed of the nullah. It was a tigress charging at full speed. Like a flash of lightning she had cleared all obstacles, and was in the first fork of Gibbon's tree eight feet from the ground, and perpendicular to it. Gibbon fired down upon her, and she fell to the earth with her jaw broken, but instantly charged again to the same spot, when another sportsman hit her with an Express bullet in the back, making a fearful wound.

"The pursuit on elephants now commenced. There were three of them, and each had a line of his own to investigate. One called Bahadur Gūj was much the stanchest, and knew what it was to be clawed.

"Just as this elephant was passing a thick spot, the wounded tigress sprang on his head. There was a brief but exciting struggle. Bahadur Gūj got his enemy down, trampled it to death, and then flung its body up on to the bank of the nullah. . . . Fortunately for the elephant, the tiger's jaw was broken, so that he received no injuries worth mentioning.

"The following incidents will show, I think, what a mistake it is to suppose that tigers are never found except in the near neighborhood of water during the hot months of the year. Whilst out with a party of four, in the middle of May, we beat unsuccessfully for a fine tigress that had killed a cow during the previous night. The beat was properly conducted, but no beast of prey appeared. A mile or two distant there was a very fine jungle, but it was decided that as there was no water, there could be no tiger in it. We therefore thought it a good opportunity to organize a beat on behalf of our native shikáris, in order that they might slay for themselves deer, pig, and such like animals for their own eating.

"Accordingly, we repaired to the desired locality, and scattered ourselves about without taking any of the usual precautions. Some of us helped in the beat, and some of the beaters converted themselves into shooters, and took up such positions as seemed good to them. Things were proceeding very pleasantly, when suddenly a shot was fired by one of the natives, and word was rapidly passed that he had aimed at a tiger, which had not fallen, but gone on up a ravine towards the head of the jungle. No blood marks were found, and the bullet was held to have missed. This was ultimately found to be true. But at the moment I doubted it, for the man was an excellent shot, and the tiger had come out slowly just in front of him.... At all events, the tiger was gone, and I and my friend had to do our best to find him. The elephant Bahadur Gūj was called up, and I and my companion stood up in front of the howdah, while the native who had first fired at the animal occupied a back seat with his little son.

"For a long time our search was fruitless. We worked up to the head of the jungle without finding a vestige of the enemy. On our way back my coadjutor pointed to a thick corinda bush and said, 'That is a likely spot.' I looked, and there was the tiger, or rather tigress, standing in the centre of it. We fired together. There was a roar, a scuffle, and a dense cloud of smoke, under cover of which the tigress disappeared, having only been seen by the small boy in the back seat. The cover consisted entirely of detached bushes, so we felt sure she could not have gone far. At last we discovered a black hole flush with the ground. This we approached cautiously, and on peering down saw the legs of a recumbent tiger. We threw stones in, but the animal never moved; and on getting a view of her head, my friend put a ball through it. Three of us now got down into the den, and with much difficulty contrived to get the beast out without injuring the skin."

Looking around once for a wounded tiger in the Nielgherries by night, Major Leveson and his party drove the beast into a patch of jungle, "not more than fifty yards long by twenty wide. Chinneah (the head shikári) threw a couple of lighted rockets into this retreat, which evidently annoyed him, although they had not the effect of causing the animal to break; it only set up a low angry growl that lasted for some time. Two or three times I saw the bushes shake as if it were about to spring; and once I caught a hurried glimpse of its outline, and threw up my rifle, but put it down again, as I did not like to fire a chance shot with an uncertain aim. Again Chinneah's rockets flew hissing about the tiger, and caused him to move, for B —— caught sight of him and let drive right and left. Then out he sprang with an appalling roar, and struck down poor Ali, who, notwithstanding my orders, had separated himself from the rest in order to pick up a stone to throw into the bush. His piercing death shriek rang through the night air, striking terror to every heart; and although I knew that it was too late to save him, I determined that he should be revenged, and dashed forward towards the spot where the infuriated brute was savagely growling as it shook the senseless but quivering body. No sooner did I get a glimpse of the tiger than I knew I was perceived, for with a short angry roar he left the corpse, and crouched low upon the ground, with head down, back arched, and tail lashing his heaving flanks. At this moment . . . carefully aiming between the eyes which glared upon me like balls of fire, I fired — he reared up at full length, and fell back dead.

"Vengeance satisfied, I went up to poor Ali, whom I found shockingly mutilated. His death must, however, have been instantaneous, as the tiger with the first blow had shattered his skull and scattered his brains about the spot."

The hunting tiger is not the highest development of his species. He has not much to learn, compared with a man-eater, in order to adjust himself to the requirements of life; and the gaunt, somewhat undersized, active, hardy, shy and solitary beast, pursues the tenor of his way far from the habitations of men, of whom he is wary and distrustful, chiefly on account of their strangeness.

To a cattle-lifter life presents more diversified scenes. The way in which the animal lives implies a greater complexity of conditions to which he is required to adapt himself, and a corresponding development of faculty. This kind of tiger, except under circumstances which rarely occur, is both a game-killer and beef-eater. Few districts yield a constant supply in the way of cattle, and when that fails, necessity compels the marauder to hunt almost exclusively, or take to homicide. On the one hand, these creatures have the experiences and training of their brethren belonging to the wastes; on the other, they are to some extent brought into a certain relationship with human beings, become accustomed to them, observe their actions, and are familiarized during those plundering expeditions, by which they mainly support themselves, with a variety of things which are altogether outside the ordinary experiences of wild beasts. Of the two classes, it goes without saying that the latter must be the more evolved; for it is not more certain that, other things being equal, the man who has had most training will be most capable, than it is that the same effects will follow in the case of tigers.

Those regions inhabited by hunting tigers have not failed to contribute, through the influence of their associations and scenery, to that vague body of feeling and of imaginative impressions, which most persons carry with them concerning this suggestive animal. "Tigers," remarks Sherwell, "are prone to haunt those crumbling works belonging to states and dynasties that have been swept away by war." In the deserted fortress of Mahoor, says Major Bevan, they were "so abundant that a few matchlock-men, who had been kept there to guard the temple, were afraid to go occasionally to the arsenal to bring their ammunition." The jungles and forests where game-killing tigers prowl for their prey are among those scenes in nature which no man who has appreciated their full significance ever forgets. "They who have never explored a primeval forest," writes Leveson, "can have but a very faint impression of the mysterious effect that absence of light and intense depth of gloom . . . the unbroken stillness and utter silence . . . exert upon the mind." They "create a strange feeling of awe and loneliness that depresses the spirits and appalls the hearts of those unaccustomed to wander in these solitudes. . . . Solitude is too insufficient a term to convey an idea of the overpowering sensation of desolation and abandonment that pervades these regions."

Stranger, perhaps stronger than all else, is the bewildering feeling of contrast between the impressive actualities of one's surroundings, and the spectral appearance of whatever the eye takes in. Peril may be imminent at every step, and yet all things seem unreal in that weird atmosphere in which they are seen. Animals look like the shadows of themselves. An elephant's motionless, gigantic form, looming even larger than in life, will define itself upon the sight, vanish as you gaze, and by some new effect of light, reappear in the same spot and the same position. It is like being in the enchanted forests of old romances; and such impressions can scarcely have failed to influence many whose exploits were performed amid such scenes. Leveson, in a place like this, saw the only encounter that has been described between the tiger and a bison bull.

"Whilst hunting in the jungle between the Bowani River, and the Goodaloor Pass, at the foot of the Nedeniallah Hills, my friend Burton and I witnessed a most gallantly-contested fight between a bull bison and a tiger. . . . Night had scarcely set in when a loud bellowing was heard, followed by an unmistakable roar that caused no little commotion amongst the horses and bullocks that were picketed round our tents. From the ominous sounds which succeeded we knew that a mortal combat was raging at no great distance from our bivouac. Having arranged for the safety of our camp, Burton and I, armed with rifles and pistols, followed closely by Chinneah and Googooloo, each carrying a couple of spare guns, sallied forth; and keeping along the bank of the river for a short distance, entered the dense cover, from which the sounds of the contest seemed to issue, by a narrow deer-run. Here we could only get along very slowly, having to separate the tangled brushwood with one hand, and hold our rifles cocked and ready with the other. We proceeded in this manner for some distance, guided by the noise of the contest, which sounded nearer and nearer, and came to an opening in the woods where we saw a huge bull bison, evidently much excited, for his eyes flashed fire, his tail was straight on end, and he was tearing up the ground with his forefeet, all the while grunting furiously. As we were all, luckily, well to leeward, the taint in the air was not likely to be winded, so I made signs to the bearers to lay down their guns, and climb into an adjacent tree; while Burton and myself, with a rifle in each hand, by dint of creeping on our hands and knees, gained a small clump of bushes on a raised bank, and not more than thirty yards distant, whence we could see all that was going on. When we first arrived, the tiger was nowhere to be seen; but from the bison's cautious movements, I knew he could not be far off. The moon was high in the heavens, making the night as clear as day; so not a movement could escape us, although we were well concealed from view.

"Several rounds had already been fought, for the game had been going on a good twenty minutes before we came up, and the bison, besides being covered with lather about the flanks, bore several severe marks of the tiger's claws on the face and shoulders. Whilst we were ensconcing ourselves comfortably behind the cover, with our rifles in readiness for self-defence only, — for we had no intention of interfering in the fair stand-up fight which had evidently been taking place, — a low savage growling about fifteen paces to the right attracted our attention; and crouched behind a tuft of fern, we discerned the shape of an immense tiger watching the movements of the bison, which, with his head kept constantly turned towards the danger, was alternately cropping the grass, and giving vent to his excited feelings every now and then by a deep, tremulous roaring, which seemed to awaken all the echoes of the surrounding woods. The tiger, whose glaring eyes were fixed upon his antagonist, now and again shifted his quarters a few paces either to the right or the left, once coming so near our ambuscade that I could almost have touched him with the muzzle of my rifle; but the wary old bull never lost sight of him for a second, and ever followed his motions with his head lowered to receive an attack. At last the tiger, which all along had been whining and growling most impatiently, stole gently forward, his belly crouching along the ground, every hair standing on end, his flanks heaving, his back arched, and his tail whisking about and lashing his sides; but before he could gather himself together for a spring which might have proved fatal, the bison, with a shriek of desperation, charged at full speed, with his head lowered and the horns pointed upward, but overshot the mark, as his antagonist adroitly shifted his ground just in time to avoid a vicious stroke from the massive horns. Then making a half circle, he sprang with the intention of alighting on the bison's broad neck and shoulders. This the bull evaded by a dexterous twist; and before his adversary could recover himself, he again rushed at him, caught him behind the shoulders with his horns, and flung him some distance, following up to repeat the move, but the tiger slunk away to gather breath.

"Round after round of the same kind followed, allowing breathing-time between each, the tiger generally getting the worst of it, for the bull sometimes received his rush on his massive forehead and horns, and threw him a considerable distance, bruised and breathless, although his skin seemed to be too tough for the points to penetrate. Once, however, I thought the bison's chance was all over, for the tiger, by a lucky spring, managed to fasten on his brawny shoulder, and I could hear the crunching sound of his teeth meeting again and again in the flesh, while the claws tore the flank like an iron rake. With a maddening scream of mingled rage and pain, the bull flung himself heavily on the ground, nearly crushing his nimble adversary to death with his ponderous weight; and the tiger, breathless and reeling with exhaustion, endeavored to slink away with his tail between his legs. But no respite was given, his relentless foe pursued with roars of vengeance, and again rolled him over before he could regain his legs to make another spring. The tiger, now fairly conquered, endeavored to beat a retreat, but this the bison would not allow. He rushed at him furiously over and over again; and at last, getting him against a bank of earth, pounded him with his forehead and horns until he lay motionless, when he sprang with his whole weight upon him, striking him with the forefeet, and displaying an agility I thought incompatible with his unwieldy appearance.

"The combat, which had lasted over a couple of hours, was now over, for the tiger, which we thought might be only stunned, gave unmistakable signs of approaching dissolution. He lay gasping, his mouth half open, exposing his rough tongue and massive yellow teeth. His eyes were fixed, convulsive struggles drew up his limbs, a quiver passed over his body, and all was still. His conqueror was standing over him with heaving flanks, and crimsoned foam flying from his widely distended nostrils; but his rolling eye was becoming dim, for the life-blood was fast ebbing from a ghastly wound in the neck, and he reeled about like a drunken man, still, however, fronting his dead enemy, and keeping his horns lowered as if to charge. From time to time he bellowed with rage, but his voice became fainter, and at last subsided into a deep hollow moan. Then his mighty strength failed him, and he could not keep on his legs, which seemed to bend slowly, causing him to plunge forward. Again he made a desperate effort to recover himself, staggered a few paces, and with a surly growl of defiance, fell never to rise again; for, after a few convulsive heavings, his body became motionless, and we knew that all was over."

How often a conflict between animals so formidable ends in the assailant's repulse or death, we do not know, neither can we say whether bisons are habitually attacked by tigers. Lions destroy the African buffalo either singly or by taking odds; and in a personal contest, the tiger would generally have the advantage over a lion. They have often been pitted against each other, and the general result is well known to be as stated. Gunga, who belonged to the King of Oude, killed thirty lions, and destroyed another after being transferred to the zoölogical garden in London.

When the young tiger first makes his appearance among the fastnesses of forests, he is one foot long, has but little coat, although his stripes can be seen, and is blind. On the eighth or tenth day his eyes open, and by that time he has grown four inches and a half. At nine months the length is five feet, and at the expiration of a year he measures five feet eight inches. When two years of age the male's length from tip to tip is about seven feet six inches, and that of the tigress seven feet. Between the second and third year they separate from their mother. While in the days of his youth the *lodia bagh* makes indiscriminate war upon the brute creation, commits unnecessary murders, stalks his prey instead of surprising it, and, Leveson and others assert, chases it like the cheetah. But time diminishes nervous energy, and leaves him, like all other beings, bereft of the incitements its excess engenders. Experience warns him against the consequences of temerity, and he grows lazy. Then these animals take to ambushing deer-runs and drinking places; they round up game by moving round and roaring; they practise upon the curiosity which besets the *Cervidæ;* and partly show themselves in the jungle to tempt an axis deer to a closer inspection; they are also said to bark in imitation of the sambur stag, in order to lure a doe or some pugnacious buck, within reach of a rush.

As for the beast that takes to man-eating, what was most probably at first an accidental event, now becomes the occupation of its life. In the first place it encountered men casually, now this is done with intention. He *must* study the habits of his game, and that he does so, is attested by his fatal success. *Admie khane wallah*, the eater of men, glares upon them from every "coign of vantage"; he discriminates between individuals, classes, and occupations, he learns the ways of farmers and woodcutters, of women who wash by the stream, of mail-carriers, and travellers on roads, of priests who serve at lonely shrines.

No country is so favorable for his exploits as India. The endless divisions of its people into castes or professions is destructive to unanimity of feeling and combined action. The "gentle Hindu," who is one of the most callous and unsympathetic of mankind, folds his hands when one of his co-religionists has been carried off, and says that Kali probably sent the tiger for that especial purpose, so what has he to do with it? His Mussulman acquaintance twists his mustache, and mutters, *Ul-humd-ul-illa*, praise be to God, this man was only an infidel, and it was his destiny! They cannot act together, and formerly matters were worse than they are now.

Nothing could suit the prowling tiger better than these isolated settlements with their careless, nearly defenceless inhabitants, the by-ways and wastes that separate them. When he has once killed a man, and has discovered the creature's feebleness, those horrors so often recorded follow as matters of course. Henceforth, nobody is safe beyond the walls of his town or dwelling. Occasionally not even there, for the man-eater combines the extremes of conduct, — excessive wariness and desperate audacity.

There is no necessity to multiply references as to the fact that these tigers are audacious, — that is generally known to be the case; but it is well to remember in connection with their relations to mankind, that they are apt to become panic-stricken at anything which appears strange and unaccountable. Colonel Pollok preserves an incident ("Sport in British Burmah") which illustrates their enterprise, and yet shows how they become confused, incapable, and appalled by whatever is beyond comprehension,—a feature in the animal's character, by the way, which is much more creditable to its intellect than derogatory to its courage.

Hill, the officer to whom the adventure happened, relates his own experiences. He was out with a body of native troops after some Shan mutineers at the time, and in a country that Crawfurd, Colonel Yule, Hallett, Colquhoun, etc., speak of as much infested by tigers. At Yonzaleem a report was brought to him that a scourge of this kind was in the neighborhood, and that fifteen men had been killed in a month; but duty called, and there was no time in which to go hunting. "We were travelling along a mountain pathway fringed with bamboo-like grass," Hill says, "and I was leading the way about thirty paces, perhaps, in front of the party, followed at a little distance by my lugelay, or Burmese boy, carrying my loaded gun. I had nothing in my hand but my oak stick, but you know what a shillelah it is, and what a thundering blow can be given with it. It was still early, and as I was trudging along carelessly, the men behind me jabbering and talking, I heard a slight noise on the edge of the pathway to my right; for a second I paid no attention to it, but thinking it might be a jungle-fowl or a pheasant, I beckoned to the boy to give me my gun. He had loitered behind, and before he could reach me, by slow degrees out came the head of an enormous tiger, close to me, almost within hitting distance. Unfortunately my lad, and the Burmese escort, saw it too, and halted, calling out 'The tiger! the tiger! he will be killed! he will be killed!' meaning me. I did not take my eyes off the tiger's, but put my hand behind my back, saying in Burmese to the boy, 'Give me my gun;' but he and the others only kept jabbering, 'He will be killed! he will be killed!' Not a man stirred, though they were all armed and loaded. So there we were, the tiger and I, face to face. At last, thinking to frighten it away, I lifted the stick and pretended to hit it a back-handed blow, at the same time making a sort of yelling noise. The stick was over my left shoulder, but so far from being intimidated, the tiger rushed at me, and I caught him a blow on the side of the head and floored him.

"Seeing him pick himself up with his back towards me, I thought he was going to bolt, and for the first time turned round, and said, 'Now give me my gun.' Before the words were well out of my mouth, my stick was sent flying, my right hand pinned to my side by one of his hind claws, and one of his fore-paws on my shoulder and back, and he stood over me growling in a most diabolical manner. I bent my back, stuck out my legs, and with my left arm struck towards my right shoulder at the brute's face, which was towering over me, snarling and growling like the very devil. Suddenly, with an infernal roar, he struck me on the neck, and down I went as if I had been shot, the tiger turning a somersault over me, and falling on his back. In a second, in my endeavors to get up, I was on my hands and knees, the blood pouring over my face, beard and chest, giving me, I have no doubt, a most satanic appearance. As the tiger recovered we met face to face. He looked at me, seemed to think that by some strange metamorphosis, from a two-legged man, whom he despised, I had become some kind of a four-legged monster like himself, put his tail between his legs, and bolted for his life."

This is a very disconcerting account for those who assert that the tiger is always dazed by daylight, and a coward at all times; that he shrinks from the sight and scent of human beings, flies from the sound of the human voice, and quails before the glance of a man's eye.

Colonel Pollok ("Natural History Notes") says he "never heard of a black tiger," but that he has "seen the skins of three white ones; two entirely white and the other faintly marked with yellow stripes." These came from the mountains of Indo-China. In the Himalayas they have been shot at an elevation of eight thousand feet above the sea, and, besides being what is called white, were maned. J. W. Atkinson ("Travels on the Upper and Lower Amoor") tells of a young Kirghis who, while carrying off his bride, camped on this river and lost her there by a tiger's attack. He threw away his own life in following this animal, dagger in hand, into the reeds. This does not always happen so by any means. Asiatics do what Europeans cannot attempt. It is well known that the Ghoorkas kill tigers with their celebrated knives; but we do not hear how many of them are destroyed in such combats. Captain Basil Hall ("Travels in India") saw a Hindu (using one of these weapons) meet a tiger at a Rajah's court, evade his spring, hamstring him as he passed, and cut through his neck into the spinal cord when the brute turned. In ancient times that class of gladiators called *Bestiarii*, encountered tigers in the Roman arena; and if one may judge from notices that are rather vague, they were pretty generally expended. The Brinjarries, says Forsyth, sometimes, assisted by their dogs, assail them with lances; and they were certainly killed by arrows at one period, but in what proportion to those whom they slew is unknown.

Certain traits are common to all the race; and as a summary of the foregoing, the appended remarks and illustrations will not be out of place. Wherever the tiger is found, water, despite Colonel Barras' solitary voice to the contrary, must be near. He drinks much and often, and cannot live in arid places. Therefore it is that the time to hunt him in India is during the hot season. Those spots where he resorts for water, and what is equally necessary to him, shade, are well known in all parts where he is to be found; and it is there that buffaloes — young ones, for an ordinarily fastidious tiger will not touch an old, tough animal — are tied up. When taken, his trail is followed to the spot where he makes his lair.

There is one exception, however, to all rules that usually govern the pursuit of tigers. When a man-eater is the object, the trailing must go on all day and every day until this monster is run down. No better example of what is to be done under these circumstances can be given than Captain Forsyth's narrative of his own exploit in the Bétúl jungle.

"I spent nearly a week . . . in the destruction of a famous man-eater, that had completely closed several roads, and was estimated to have devoured over a hundred human beings. One of these roads was the main outlet from the Bétúl teak forests, towards the railway under construction in the Harbadá valley; and the work of the sleeper-contractors was completely at a stand-still, owing to the ravages of this brute. He occupied regularly a large triangle of country between the rivers Móran and Ganjál; occasionally making a tour of destruction much further to the east and west, and striking terror into a breadth of not less than thirty or forty miles. It was therefore supposed that the devastation was caused by more than one animal; and we thought we had disposed of one of these early in April, when we killed a very cunning old tiger of evil repute after several days' severe hunting. But I am now certain that the one I destroyed subsequently was the real malefactor, since killing again commenced after we had left, and all loss of human life did not cease till the day I finally disposed of him.

"He had not been heard of for a week or two when I came into his country, and pitched my camp in a splendid mango grove near the large village of Lokartalae, on the Móran River.

"A few days of lazy existence in this microcosm of a grove passed not unpleasantly. . . . In the mean time I was regaled with stories of the man-eater — of his fearful size and appearance, with belly pendent to the ground, and white moon on the top of his forehead; his pork-butcher-like method of detaining a party of travellers while he rolled himself in the sand, and at last came up and inspected them all round, selecting the fattest; his power of transforming himself into an innocent-looking woodcutter, and calling or whistling through the jungle till an unsuspecting victim approached; how the spirits of all his victims rode with him upon his head, warning him of every danger, and guiding him to the fatal ambush where a traveller would shortly pass. All the best shikáris of the country-side were collected in my camp, and the landholders and many of the people besieged my tent morning and evening. The infant of a woman who had been carried away while drawing water at a well was brought and held up before me, and every offer of assistance in destroying the monster made. No useful help was, however, to be expected from a terror-stricken population like this. They lived in barricaded houses, and only stirred out, when necessity compelled, in large bodies, covered by armed men, and beating drums and shouting as they passed along the roads. Many villages had been utterly deserted, and the country was being slowly depopulated by a single animal. So far as I could learn, he had been killing alone for about a year — another tiger that had assisted him in his fell occupation having been shot the previous hot weather. Bétúl has always been unusually afflicted with man-eaters, the cause apparently being the great numbers of cattle that come for a limited season to graze in that country, and a scarcity of other prey at the time when these are absent, combined with the unusually convenient cover for tigers alongside of most of the roads. The man-eaters of the Central Provinces rarely confine themselves *solely* to human food, though some have almost done so to my own knowledge.

"As soon as I could ride in the howdah [Captain Forsyth was suffering from an accident at this time], and long before I was able to do more than hobble on foot, I marched to a place called Chárkhérá, where the last kill had been reported. My usually straggling following was now compressed into a close body, preceded and followed by baggage-elephants, and protected by a guard of police with muskets, peons with my spare guns, and a whole posse of matchlock shikáris. Two deserted villages were passed on the road, and heaps of stones at intervals showed where some traveller had been struck down. A better hunting-ground for a man-eater certainly could not be found. Thick, scrubby teak jungle closed in the road on both sides; and alongside of it for a great part of the way wound a narrow, deep watercourse, overshadowed by jámare bushes, and with here and there a small pool of water still left. I hunted along this nálá the whole way, and found many old tracks of a very large male tiger, which the shikáris declared to be those of the man-eater. There were none more recent, however, than several days. Chárkhérá was also deserted on account of the tiger, and there was no shade to speak of; but it was the most central place within reach of the usual haunts of the brute, so I encamped there, and sent the baggage-elephants back to fetch provisions. In the evening I was startled by a messenger from a place called Lá, on the Móran River, nearly in the direction I had come from, who said that one of a party of pilgrims who had been travelling unsuspectingly by a jungle road, had been carried off by the tiger close to that place. Early next morning I started off with two elephants, and arrived at the spot about eight o'clock. The man had been struck down where a small ravine leading to the Móran crosses a lonely pathway a few miles east of Lá. The shoulder-stick with its pendant baskets, in which the holy water from his place of pilgrimage had been carried by the hapless man, was lying on the ground in a dried-up pool of blood, and shreds of his clothes adhered to the bushes where he had been dragged down into the bed of the nálá.

"We tracked the man-eater and his prey into a very thick grass cover, alive with spotted deer, where he had broken up and devoured the greater part of the body. Some bones and shreds of flesh, and the skull, hands, and feet were all that remained. This tiger never returned to his victim a second time, so it was useless to found any scheme for killing him on that expectation. We took up his tracks, however, from the body, and carried them patiently down through very dense jungle to the banks of the Móran,— the trackers working in fear and trembling under the trunk of my elephant, and covered by my rifle at full cock. At the river the pugs [footprints] went out to a long spit of sand that projected into the water, where the man-eater had drunk, and then returned to a great mass of piled-up rocks at the bottom of a precipitous bank, full of caverns and recesses. This we searched with stones and some fireworks I had in the howdah, but put out nothing but a scraggy hyena, which was, of course, allowed to escape. We searched about here all day in vain, and it was not till nearly sunset that I turned and made for camp.

"It was almost dusk, when we were a few miles from home, passing along the road we had marched by the former day, and the same by which we had come out in the morning, when one of the men who was walking behind the elephant started and called a halt. He had seen the footprint of a tiger. The elephant's tread had partly obliterated it, but further on where we had not yet gone it was plain enough, — the great square pug of the man-eater we had been looking for all day! He was on before us, and must have passed since we came out in the morning, for his track had covered that of the elephants as they came. It was too late to hope to find him that evening, and we could only proceed slowly along on the track, which held to the pathway, keeping a bright lookout. The Lállá [Forsyth's famous tiger-hunting shikári] indeed proposed that he should go on a little ahead as a bait for the tiger, while I covered him from the elephant with my rifle. But he wound up by expressing a doubt whether his skinny corporation would be a sufficient attraction, and suggested that a plump young policeman, who had taken advantage of our protection to make his official visit to the scene of the last kill, should be substituted — whereat there was a general but not very hearty grin. The subject was too sore a one in that neighborhood just then. About a mile from the camp the track turned off into a deep nálá that bordered the road. It was now almost dark, so we went on to camp, and fortified it by posting the three elephants on different sides, and lighting roaring fires between. Once during the night an elephant started out of its deep sleep and trumpeted shrilly, but in the morning we could find no tracks of the tiger near us. I went out early next morning to beat up the nálá, for a man-eater is not like common tigers, and must be sought for morning, noon, and night. But I found no tracks save in the one place where he had crossed the ravine the evening before, and gone off into thick jungle.

"On my return to camp, just as I was sitting down to breakfast, some Banjárás [carriers, and probably gypsies] from a place called Décknà — about a mile and a half from our camp — came running in to say that one of their companions had been taken out of the middle of their drove of bullocks by the tiger, just as they were starting from their night's encampment. The elephant had not been unharnessed, and securing some food and a bottle of claret, I was not two minutes in getting under way again. The edge of a low savanna, covered with long grass and intersected by a nálá, was the scene of this last assassination, and a broad trail of crushed-down grass showed where the body had been dragged down to the nálá. No tracking was required. It was all horribly plain, and the trail did not lead quite into the ravine, which had steep sides, but turned and went alongside of it into some very long grass reaching nearly up to the howdah. Here Sarjú Parshád, a large government mukna [tuskless male elephant] I was then riding, kicked violently at the ground and trumpeted, and immediately the long grass began to wave ahead. We pushed on at full speed, stepping as we went over the ghastly half-eaten body of the Banjárá. But the cover was dreadfully thick, and though I caught a glimpse of a yellow object as it jumped down into the nálá, it was not in time to fire. It was some little time before we could get the elephant down the bank and follow the broad plain footsteps of the monster, now evidently going at a swinging trot. He kept on in the nálá for about a mile, and then took to the grass again; but it was not so long here, and we could make out the trail from the howdah. Presently, however, it led into rough, stony ground, and the tracking became more difficult. He was evidently full of go, and would carry us far; so I sent back for more trackers, and orders to send a small tent across to a hamlet on the banks of the Ganjál, towards which he seemed to be making. All that day we followed the trail through an exceedingly difficult country, patiently working out print by print, but without having been gratified by a sight of his brindled hide. Several of the local shikáris were admirable trackers, and we carried the line down to within about a mile of the river, where a dense, thorny cover began, through which no one could follow a tiger.

"We slept that night at the little village, and early next morning made a long cast ahead, proceeding at once to the river, where we soon hit upon the track leading straight down its sandy bed. There were some strong covers reported in the river-bed some miles ahead, near the large village of Bhádúgaon, so I sent back to order the tent over there. The track was crossed in this river by several others, but was easily distinguished from all by its superior size. It had also a peculiar drag of the toe of one hind foot, which the people knew and attributed to a wound he had received some months before from a shikári's matchlock. There was thus no doubt that we were behind the man-eater; and I determined to follow him while I could hold out, and we could keep the trail. It led right into a very dense cover of jáman and tamarisk in the bed and on the banks of the river, a few miles above Bhádúgaon. Having been hard pushed the previous day, we hoped that he might lie up here; and, indeed, there was no other place he could well go to for water and shade. So we circled round the outside of the cover, and finding no track leading outside, considered him fairly ringed. We then went over to the village for breakfast, intending to return in the heat of the day.

"About eleven o'clock we again faced the scorching hot wind, and made silently for the cover where the man-eater lay. I surrounded it with scouts on trees, and posted a pad-elephant at the only point where he could easily get up the high bank and make off, and then pushed old Sarjú slowly and carefully through the cover. Peafowl rose in numbers from every bush as we advanced, and a few hares and other small animals bolted out at the edges — such thick green covers being the midday resort of all the life in the neighborhood in the hot weather. About its centre the jungle was extremely thick, and the bottom was cut up into a number of parallel water-channels among the strong roots and overhanging branches of the tamarisk.

"Here the elephant paused and began to kick the earth, and to utter the low tremulous sound by which some of these animals denote the close presence of a tiger. We peered all about with beatings of the heart; and at last the mahout, who was lower down on the elephant's neck, said he saw him lying beneath a thick Jáman bush. We had some stones in the howdah, and I made the Lállá, who was behind me in the back seat, pitch one into the bush.

Instantly the tiger started up with a short roar and galloped off through the jungle. I gave him right and left at once, which told loudly; but he went on till he saw the pad-elephant blocking the road he meant to escape by, and then he turned and charged back at me with horrible roars. It was very difficult to see him among the crashing bushes, and he was within twenty yards before I fired again. This dropped him into one of the channels, but he picked himself up, and came on as savagely, though more slowly, than before. I was now in the act of covering him with the large shell rifle, when suddenly Sarjú spun round, and I found myself looking the opposite way, while a worrying sound behind me, and the frantic movements of the elephant, told me I had a fellow-passenger on board I might well have dispensed with. All I could do in the way of holding on barely sufficed to prevent myself and guns from being pitched out; and it was some time before Sarjú, finding he could not kick him off, paused to think what he would do next. I seized that placid interval to lean over behind and put the muzzle of my rifle to the tiger's head, blowing it into fifty pieces with the large shell."

In Assam and other parts of Indo-China, and in the interior of Malacca, the natives are treated by tigers much after the same manner as those of India were in the days before modern inventions had modified the views of these brutes upon mankind.

A pit is an effectual device for taking tigers, but most descriptions of the way in which it is arranged are evidently incorrect. Malays, however, procure most of the animals they export by means of pits, which are constructed after the manner of those *oubliettes* or "dungeons of the forgotten," where in the good old times captives were placed who had no hope of release.

What is the tiger's temper? Conventionally, and according to common misapprehension, he is the furious and insatiable savage that Buffon paints — "*sa ferocité n'est comparable à rien.*" He is full of base wickedness and inappeasable cruelty, loves blood and carnage for their own sake, and longs continually to fly at unfortunate creatures with that *tremendæ velocitatis* of which Pliny speaks.

> "What immortal hand or eye,
> Framed thy matchless symmetry?
> In what distant deeps or skies,
> Burned that fire within thine eyes?"

writes William Blake, and then he asks, "Did He who made the lamb make thee?" The French naturalist and English poet looked at the subject from the same standpoint. It was not necessarily seen wrongly on that account, but it happened that the view taken by both was an imperfect one. Deeper insight or more profound research would have resolved uncertainty in the one case, and checked extravagance in the other. Had they read the runes of nature aright, the answer to such questionings, the rebuke to such exaggerations, would have been found stamped upon the organization of everything that lives. Physical constitution is never an accident or a mistake; it is at once the consequence of special modes of existence, and the cause of their continuance. Bodily conformation and its correlates in mental structure are to brutes absolutely determinative.

"Most carnivorous of the carnivora," writes W. N. Lockington ("Riverside Natural History"), "formed to devour, with every offensive weapon specialized to the utmost, the *Felidæ*, whether large or small, are relatively to their size the fiercest, strongest, and most terrible of beasts." The tiger stands at their head. He must needs appreciate his destructive power and feel the desire to exercise it. Inherited tendencies and the pressure of necessity put his capabilities into action. Their exercise, transmitted traits, and those experiences implied in habit, make him what he is, — audacious, treacherous, wary, cunning, ferocious. These characteristics answer to the anatomical specialties by which his frame is distinguished, — his convoluted and back-reaching forebrain, protective coloring, differentiated and perfectly innervated muscles, his simple digestive tract, formidable armature, and padded feet.

THE PUMA.

THE PUMA

WHAT is true with regard to the present geographical distribution of the cats, has been true always; throughout their fossil history the greater and more formidable *Felidæ* have been confined to the Eastern Hemisphere. A number of American species exist, however, ranging from among the smallest and most beautiful forms contained in this family, up to animals that in destructive power, only give place to their great African and Asiatic allies. The puma and jaguar have not filled so large a space in zoölogical literature as the lion and tiger; they have not attracted so much general attention, and are less known. But this is, to a considerable degree, the result of accident. For the most part, those who encountered them were men of a different stamp from the famous hunters whose adventures in Asia and Africa have made the animals of their forests and plains familiar and full of interest to so large a portion of the public in civilized lands.

It is seldom that the throngs that pass before cages in which wild beasts are confined, contain a spectator who knows how perfect a creature a cat is. As a class these forms are adjusted to their place in nature better than other creatures, and also much better than the human race. Their distinctive characteristics are all strongly marked, and have persisted from a period so incalculably remote, that the *Felidæ* may in this respect be said to stand by themselves. "We have as yet," remarks A. R. Wallace ("Geographical Distribution of Animals"), "made little approach towards discovering 'their origin,' since the oldest forms yet found are typical and highly specialized representatives of a group which is itself the most specialized of the carnivora." No one acquainted with the evidence upon which this statement rests is likely to gainsay it, and its meaning is not obscure. The fact carries with it a necessary implication that animals of the species referred to, having followed a definite way of life longer than the rest, are more fit in every way to meet its requirements.

Perhaps the most striking illustration that could be given of the reality of what has been said, is the small difference actually existing between wild and domesticated cats. Domestication is so great and radical a change from the feral state, that the entire constitution of an animal is affected, — mind and body, temper, intelligence, form, color, fertility and physical capacity, are all modified. But it is not thorough enough to do away with the traits engendered in the *Felidæ*, and therefore it happens that after thousands of years, the house cat varies from the wild one so little in important and distinctive characteristics. Cattle and sheep were domesticated before the dispersion of the Aryan tribes; linguistic evidence places that fact beyond question. Cats, however, though introduced into Europe from Asia, as was the case also with the horse, ass, and goat, were no doubt first reclaimed from savage life in Egypt. On the Lower Nile domestic cats were sacred to Pasht, whom the Greeks called Bubastis, and identified with Artemis. She was represented with the head of a cat or lioness, as was Sechet also, a divinity equivalent to the Phœnician Astarte.

These personifications were not meaningless. Bast or Sechet was the patroness of the baser passions and more destructive vices. It was her part, likewise, to torture the condemned in the lower world. Naturalists (Pastophori) belonging to the faculties established at "the hall of the ancients" in Heliopolis, and "the house of Seti" in Thebes, knew much more, and also much less, about zoölogy and its allied sciences than is popularly supposed.

Felis concolor, the puma, cougar, panther, mountain lion, etc., is more correctly called by the last of these names than by that of panther, under which he is commonly known throughout the northern part of this continent. In its habits the puma is said, but not with any great degree of appropriateness, to resemble the leopard more closely than any feline species. Buffon called it the American lion, but he knew very little about this animal, and his opinion upon its character is of no special importance. E. F. im Thurn ("Among the Indians of Guiana") remarks that in the southern part of America, and particularly in Guiana, all varieties of feral cats take their titles from the kind of game upon which they principally subsist. Thus *Felis concolor* is called "the deer tiger," *Felis nigra* the "tapir tiger," and *Felis macnera* the "peccary tiger." Such may be the case when aborigines are forced to particularize; but in common parlance one hears only the sobriquet "*leòn*" bestowed by all classes of people on the puma.

There is but one true species found in America, and this is distributed in all parts of the continent. The average length from tip to tip may be given at about six and a half feet. In maturity the skin is of a uniform tawny hue on the back and sides, with some deepening of shade in the case of individuals. Cubs are born with dark stripes upon the body, and spots on the neck and shoulders. Garcilasso de la Vega ("Royal Commentaries") speaking of this beast as the tutelar of certain noble Peruvian families, and probably their eponymous ancestor, says: "A Spaniard whom I knew killed a large lioness (female puma) in the country of the Antis, near Cuzco. She had climbed into a high tree, and was slain by four thrusts of a lance. There were two whelps in her body *which were sons of a tiger* (jaguar), for their skins were marked with the sire's spots."

Like all *Felidæ* except the cheeta or hunting leopard, the limbs have little free play; they are not adapted to continued rapid locomotion, being short and massive, very powerful, but somewhat limited in variety of action, and more capable of extreme and spasmodic efforts than of persistent use. The animal is very arboreal in its habits, and its climbing powers and general dexterity are not surpassed by any species belonging to this family.

Like true panthers, these cougars, carcajous, catamounts or pumas (the native title is *sassu-arana* or false deer) are, according to H. W. Bates ("The Naturalist on the River

Amazon "), accustomed to live in cliffs and caves, and they seem able to do without the constant supply of water that some others among the *Felidæ* require.

It is said that here, as in India, the representatives of the tiger and lion do not live together. While this may be true in a general way, there is not the same separateness of range as in Asia; and the author, in common with other explorers, has found them in similar localities on several occasions. No accounts have been given, so far as the present writer is aware, of actual conflicts occurring between the puma and jaguar, and, in fact, there could be little hope for the former in such a contest, as his adversary would be much heavier and more powerful, equally active, and better armed. With respect to the grizzly bear, there is little doubt that common report among frontiersmen, to the effect that he is often assailed by the puma and frequently worsted, has some foundation in fact. From two to four young are born together, and by the end of the first year these whelps lose their spots and stripes. They are lively and playful during infancy, and although in them, as in all animals so highly organized, a decided individuality displays itself from the first, personal experience has convinced the author that they possess a great degree of intelligence, are easily taught those things which their faculties enable them to acquire; and, so far as their own interest and convenience influence conduct, that they exhibit ludicrously strong preferences and dislikes.

Great strength and activity are combined in the puma, its armature is formidable, the brute is habitually silent, stealthy in the highest degree, and full of the so-called treachery of its race. Besides this, it is very enterprising when occasion warrants a display of audacity, as well as extremely ferocious and blood-thirsty. More frequently, perhaps, than any of the great cats, it kills for the mere gratification of its cruel impulses. Dr. Merriman ("Mammals of the Adirondacks") states that on level ground " a single spring of twenty feet is not uncommon for a cougar," and Sheppard records the measurement of a distance twice as great when the leap was made downward from a ledge of rock upon a deer.

Padre José de Acosta ("Historia natural y moral de las Indias") says that neither the puma nor the jaguar "is so fierce as he appears to be in pictures," though both will kill men. There are, however, many places where the puma has been so cowed by ill success in his attacks upon human beings, that he avoids them as much as possible. Cieza de Leon and Garcilasso de la Vega express themselves to the same effect. Humboldt found whole villages abandoned by their helpless inhabitants in consequence of the ravages of the two great American cats, but Emmanuel Liais ("Climats, Géologie, Faune, etc. du Brésil") asserts that both "*l'une et d'autre fuient l'homme et les chiens; même un enfant à cheval leur fait peur.*" This is a mere repetition of what has been asserted without qualification, proper inquiry, or adequate experience with the larger *Felidæ* in Asia and Africa.

There is no need to argue the question whether or not pumas can or will kill men; that has been affirmatively settled by facts. This creature's personal courage is a different matter. It is only a brute; yet if any one studies what has been said with regard to this trait, it will appear that most denunciations of the animal's cowardice rest upon circumstances under which it did not conduct itself like a gentleman. A cougar's padded foot, its short massive limbs, which prevent it from chasing prey, the brute's great powers of concealing itself, and perfect physical adjustment to sudden and violent attacks, are recapitulated as though they had no necessary connection with its behavior, and were not inseparably associated with corresponding peculiarities of character and habit.

A beast of prey passes the active portion of its existence in projecting or executing acts of violence. Habitual success means life, and failure death. Under such circumstances, under the influence of an experience in which by far the larger part of those enterprises undertaken resulted favorably, a self-confidence, incompatible with cowardice, will ensue.

At the same time there seems to be some general preconception with respect to the character of wild beasts, such as converts every manifestation of prudence into poltroonery. The clash of opinions expressed about all the more imposing animals witnesses to the crude and arbitrary manner in which they have been formed. With respect to this one, not the tiger himself has been the subject of more irreconcilable statements.

Stories of puma hunting and of the animal's exploits depend, so far as their style is concerned, upon the place where they are told, and the experiences of the narrator. No hunter of large game thinks it anything of a feat to shoot a cougar, yet the author has known these brutes to fight desperately when brought to bay, and in two instances their resistance was sufficiently formidable to cause, in the one case loss of life, and in the other injuries from which men never entirely recovered. Many such examples might be gathered, but they are nevertheless exceptional. A puma is not difficult to kill, and if it is seen in time, a properly armed man must either be very unfortunate or very unfit for the position in which he finds himself, if the result is not favorable. What is said of the panther and leopard, however, by Captain Forsyth ("The Highlands of Central India") and by Sir Samuel Baker ("Wild Beasts and their Ways") is peculiarly applicable to this animal: it is almost always met with unexpectedly, and no mortal can say beforehand what it will do. If taken at advantage and by surprise, as commonly happens, a single man would not usually have much chance at close quarters. The writer has, however, known them to be killed with knives, though not without severe injury to the victor.

The average native of tropical America, while fully appreciating how much more dangerous is the beast he calls a tiger, is quite enough impressed with the prowess of its smaller, though sometimes equally ferocious ally, to have his mind saturated with superstitions concerning pumas. Tapuyo or Mameluco guides will sit by a camp fire and talk in a way to put Acuna or Articda in the background. Almost equally with the jaguar this creature has supernatural and diabolic connections. When its rarely heard cry or scream, as any one may choose to call a sound so difficult to describe and which varies so greatly, floats through the forest, these natives never know whether they hear a prowling cougar, or the voice of that god from whom its race descended. Botos, a demon of woodland lakes, guides the beast to his prey; the basilisk worm Minhocao is somehow connected with it in its designs against human beings, and the deadly man-like Cæpora shrieks in concert with pumas as they roam through the darkness. W. A. Parry ("The Cougar") says that its cry "can only be likened to a scream of demoniac laughter," and that the female's answer to her mate's call resembles "the wail of a child in terrible pain."

James Orton and Prince Maximilian of Nieuwied have severally settled it that cougars are all abject cowards. Speaking from personal recollection, the author feels no hesitation in saying that it required great singleness of mind to come to this conclusion, and much dexterity to go where they did and avoid seeing things which might have modified this conclusion.

It does not follow, for reasons which have been explained at length, that because a puma attacks a grizzly bear he must be dangerous to a man; or because numbers of men have undoubtedly been killed in some places, that it should be formidable to human beings everywhere.

"When hungry," says Theodore Roosevelt ("Hunting Trips of a Ranchman"), "a cougar will attack anything it can master." Audubon, however, supposes that it never ventures to assail such large animals as cows or steers. William B. Stevenson ("Twenty Years in South America") tells us how destructive this creature is to horses, and also how the more than half-wild cattle of the pampas form into rings to defend themselves. Captain Flack ("A Hunter's Experience in the Southern States of America") relates an incident in which his horse was stalked by a cougar. S. S. Hill ("Travels in Peru and Mexico") informs us that "this animal always flies at the sight of man." G. W. Webber ("The Hunter Naturalist") declares that he "knows hundreds of well-authenticated instances in which the cougar or panther attacked the early hunters — springing suddenly upon them from an ambush." Many writers affirm that calves, colts, sheep, goats, swine, are the only domestic animals ever preyed upon, and a deer the largest wild creature which is destroyed. But a traveller like Charles Darwin was certain to observe that, although in La Plata "cougars seldom assault cattle or horses, and most rarely man," living principally on ostriches, deer, bizcacha, etc., in Chile, they killed all those animals they are said never to touch, including man.

Moreover, we read dogmatic assertions to the effect that pumas always leap on their victims from behind, and break their necks by bending back the head. Another authority decides that this is so far from being the case that death commonly arises from dislocation caused by a blow with the paw; still another insists that the vertebræ are not disjointed at all, but bitten through, which is again denied by those who are convinced that cougars invariably kill their prey by cutting the throat. Much the same statements are made about everything the beast does or is said to do, and the conclusion, which one familiar with this kind of literature comes to, is that these conflicting statements are not all false, but in a restricted sense all true. That is to say, the several ways of destruction mentioned are practised as occasion requires or suggests.

One point at least with regard to the puma's disposition in certain directions is more clearly set forth than has been the case in respect to other beasts of prey, and this is the fact that the creature's temper has been greatly changed by contact with mankind. The same thing has happened everywhere with all game hunted successfully for a long period; but this fact is ignored, and brutes whose natures are different in some minor traits from what they once were, are discussed as if the special features now exhibited had been always the same.

C. Barrington Brown ("Canoe and Camp Life in British Guiana") relates an incident which occurred while he was exploring the upper courses of the Cutari and Aramatau rivers. "One evening, while returning to camp along the portage path that we were cutting at Wonobobo Falls, I walked faster than the men, and got some two hundred yards in advance of them. As I rose the slope of an uneven piece of ground, I saw a large puma (*Felis concolor*) advancing towards me, along the other side of the rise, with its nose close to the ground. The moment I saw it I stopped, and at the same instant it tossed up its head, and seeing me also, came to a stand. With its body half-crouched, its head erect, and its eyes round and black from the expansion of their pupils in the dusky light, it was at once a noble and appalling sight. I glanced back along our wide path to see if any of the men were coming, as at that moment I felt that it was not well to be alone without some weapon of defence, and I knew that one of them had a gun; but nothing could be seen. As long as I did not move the puma remained motionless also; and thus we stood, some fifteen yards apart, eyeing each other curiously. I had heard that the human voice was potent in scaring most wild beasts, and feeling that the time had arrived for doing something desperate, I waved my arms in the air and shouted loudly. The effect on the animal was electrical; it turned quickly to one side, and in two bounds was lost in the forest."

Now why did this brute thus behave? The narrative gives not the least explanation of its conduct. Brown thought it was frightened by his gestures, because a few days before he had come upon a jaguar basking on a rock by the river, whose serenity was not at all disturbed by the voices of a boat full of men. But that was merely a guess. Very probably this animal had never seen a man previously, and almost certainly not a white man in civilized costume. There was then the profound impressiveness of absolute strangeness in the sight, and this alone would have been more likely to alarm a human being or intelligent brute than any other cause we know of. Perhaps the puma had just devoured a peccary and was gorged; or possibly its keen senses revealed the approach of Brown's party, who in fact appeared almost immediately. One may see in a narrative like this, which is a fair specimen of those relations from which most dogmatic conclusions upon the character of wild beasts have been drawn, how arbitrary and unjustifiable they generally are.

Roosevelt states that a slave on his father's plantation in

Florida was passing through a swamp one night, when he was attacked by a puma. The negro was "a man of colossal size and fierce and determined temper." Moreover, he carried one of the heavy knives that are used in cutting cane. Both parties were killed after a long and desperate struggle, whose traces were plainly impressed upon the spot. But here it appears that a man was assailed, and that the beast continued its attempts to kill him after discovering that he was armed, and persisted in its attack as long as life lasted.

One evening as the author was riding towards a hacienda in Sinaloa, and was about half a league distant from it, a girl rushed to the edge of a thicket and began to scream for help. Galloping up, it appeared she had just discovered the body of her father, killed apparently by a puma, who lay dead beside him. Life was not extinct, however, although he was very badly wounded. He said that while passing, the bellowing of an ox, mingled with the cries of some kind of beast, induced him to make his way to the scene of action. There he found a large lion, as he called it, engaged in a fight with a steer, whom he had injured severely, and who was rapidly losing blood. As soon as the man appeared, the beast left the ox and made at him. There was scant time to roll his *scrape* around his left arm, and draw the long knife which every ranchero wears in the *bota* on his right leg, before he found himself in deadly conflict.

In these three anecdotes we have a very clear refutation by facts of several points with regard to this brute's character, which have been generally accepted as settled.

Wariness and an entire absence of all the sentiments that produce recklessness in man, are as distinctly marked characters among the *Felidæ* as their peculiar dentition or retractile claws; yet the author was informed by Colonel W. H. Harness that last summer (1893) a very large panther, as the animal is called in West Virginia, walked into an extensive logging camp near the town of Davis at midday; traversed one wing of the long building in which the men employed slept, and without making any demonstration of hostility towards those who fled before him, entered their dining-room and helped himself to the meat on the table; after which he quietly passed out of a side door, and was shot from a window. If this beast had been broken down with age or disabled by accident so that it could not hunt, or if the season and weather had been such as to banish game from the vicinity, its conduct might be comprehensible. This happened with an animal in perfect physical condition, and at a season when the mountains were full of game. The brute also must necessarily have connected all the men it knew anything about with death-dealing firearms, and that it then should have walked into a crowd, and lost its life in this act of seemingly idiotic bravado, simply sets at naught everything that is known of the creature's character and habits.

Pumas, like Asiatic panthers, are easily caught in traps, but independently of this form of incapacity, they are far from being wanting in sagacity. Cougars are most accomplished hunters, and it has been explained how much that means. One of them, for example, will sometimes trail a human being for a day's journey without finding what it considers to be a suitable opportunity for making an attack.

The best and most intimate acquaintance with the character of a wild beast comes from those associations involved in domestication. When you have brought up an animal and been with it constantly day by day, the chances of finding out what it is like are better than they could be under any other conditions whatever. Prince Maximilian of Nieuwied, states that the puma is "peculiarly susceptible of domestication." It does not appear, however, that he made any experiments in this direction, and it may be suspected that if he had, certain reasons for modifying his views upon the animal's character would have suggested themselves during their course. A cougar is a cat, and in virtue of that fact is, as has been said, of all animals the least susceptible of radical change. Sanderson and Barras make a wide distinction between feline species, considered as amenable or refractory to such influences; and nothing is offered in the way of disparagement to their opinions, provided it be admitted that a young tiger may be a much more amiable and interesting infant than a panther cub, and, according to Gérard, a lion whelp attaches all hearts by its good qualities. But there soon comes a time at which traits inherent in them all are developed, and when they become strikingly alike in all their essential characteristics.

The writer bears in affectionate remembrance a pet "panther" who, from earliest life until his complete and splendid maturity, lived with him upon terms of the closest companionship. Every one who seriously studies anecdotes of brute intelligence and character must necessarily distrust them. Their authors always, either directly or by implication, put inference in the place of observation, or they start with a hypothesis, the tendency of which is to assimilate evidence, and often, no doubt unconsciously, fit facts to their own preconceptions. It is hoped that the records of daily observation here made use of for the purpose of sketching traits of character, may not prove to be without some interest and value, and that their fragmentary and incomplete form will witness to the fact that nothing is given which seemed to be either speculative or unauthentic.

One sultry morning as the author sat at ease in his sala, an Indian entered and said he had heard that the Señor delighted in wild beasts, so that having by the help of God, some saints, and several friends, slain the mother of this little lion in the Golden Mountains, he had brought it there as a mark of respect, and would like to have seven Spanish dollars. Here he unrolled his *serape* and deposited a ball of indistinctly striped and spotted fur upon the floor. In that manner this puma of pumas came into the keeping of his guardian.

The latter impressed with a sense of the responsibilities attaching to the position in which he was placed, at once sprinkled the cub with red wine and called it Gato, — a procedure it resented as if the spirit of Constantine Capronymus himself had entered into its sinful little body. The rage of infancy, however, does not endure, and Gato shortly "serened himself," to use the idiom of the country, where these things took place. He inspected his new acquaintance, rubbed up against him, had his head scratched with much complacency, and graciously ate as much as he could hold. Thus we made friends, and the compact was ever after kept by both parties, each in his own way.

The panther's way was a very simple one. It consisted in looking to the being he had come in contact with for everything he wanted, and resolutely refusing to enter into intimate communications with any one else. Nobody who knew him could say that the least feeling of affection ever warmed his heart, but it was plain enough that while he contemned the human race, one man was tolerated, and a distinction made between him and all others. Some individuals he detested at first sight, and resented the slightest approach to familiarity. For the remainder he entertained a quiet contempt; but as for fearing them, nothing was further from his thoughts. So far as that went, it is very doubtful whether he ever felt any real dread of his guardian. Some feeling akin to respect may have existed in his mind. His powers of observation were keen and quick, he saw that this particular person differed in appearance from those about him, acted differently, and was somehow or other not the same as they. If he got into difficulties, and was likely to suffer the consequences of misconduct, hostilities against him ceased when his friend appeared upon the scene; he understood this perfectly, and took refuge with him when danger threatened. As was said, Gato had no affectionate impulses so far as could be certainly known. When he wanted to be stroked, or was hungry, or wished to play, or felt insecure, he came to his guardian, followed him about, and lay beside him. Moreover, the little savage was jealous. If he beheld a dog it always put him in a passion to see it coming towards his master to be caressed. He would fly to get ahead, dance about, jump on his knee, and growl and show his teeth with every sign of anger against the intruder upon his rights.

Colonel Julius Barras ("India and Tiger Hunting") speaks of the jealousy shown by tiger cubs in his possession, but whereas he was satisfied that this was an expression of tenderness towards himself, the writer thinks it more likely to have been an exhibition of selfishness. Gato manifested at a very early age an appreciation of his own possessions, and a determination to do things after his own fashion. So far from checking this by force, his guardian encouraged it, and after having come to a clear understanding with him on the subject of biting and clawing, left him alone to follow his own devices. He was a very sagacious personage, and there was not a drop of cowardly blood in his whole body. When he was a baby there was little to distinguish him, while at rest, from some domestic cats, but he no sooner began to move about than his free wild air, the unmistakable style of savagery that stamped every action, showed him in another way. It may be added that, being left free to exhibit his individuality, and not having his family and personal characteristics marred or masked by enforced restraint until the creature grew dull, apathetic, and half imbecile, he was as pretty a specimen of feline peculiarity as any one could expect to see. Nothing was clearer to him than that the many-colored rug he was accustomed to lie on was his own. He had favorite places in which to sleep, meditate, and make observations. It would have been disagreeable indeed for any servant about the establishment to take off his bright silver collar after he grew to any size, and when he captured anything and put it away, that article became his private property, and he had no notion of giving it up.

Candor compels the admission that flattering as would have been some tokens of disinterested affection, he never gave any. What he did was to please himself. When he had no desire to be taught, which was often the case, a more stupid, sulky, and unsatisfactory pupil could not be imagined; but when his interest happened to be excited he was quickness itself, and he seldom forgot. One might as well have caressed a stuffed cat, or tried to romp with a dead one, as to have expected any recognition of advances in these directions when Señor Gato felt disposed to contemplation, and if compelled, as of course was the case sometimes, to do anything against his humor, he was not accustomed to leave any doubt about the disgust and anger which possessed him. From first to last, always, and under all circumstances, he like Richard, was "himself alone," and never stooped to the snobbishness of pretence. Thus it happened that although under fostering care and paternal rule the creature grew in grace continually, he never became fitted to adorn general society. The asperity of his nature easily showed itself; the wild beast broke through the habitual dignity of his demeanor on small provocation. Not even that to him, extraordinary person with whom he was most intimate, and whose resources so powerfully impressed his mind, might pull his ears or twist his tail after he grew up. This was to pass the proper limits of familiarity, and whenever it happened he crouched and glared with glistening fangs. That was all, however; no act of hostility followed.

Gato began to stalk his guardian at an early age, but soon learned that a statue of St. John the Evangelist was not alive, and gave up his practices against the Apostle. He discovered likewise the illusory character of shadows, which at one time were taken to be substantialities, and somehow or other satisfied his mind about his own reflection in a fountain when the wind ruffled its surface. This gave him much concern for a while. Being accustomed to look at himself in a glass, and to stand with his forepaws on the edge of the basin and see his reflection in still water, what perplexed and excited him was the fact that it sometimes looked as if it moved while he was motionless. Whether he found out about the ripple, nobody knows, but he stopped tearing round the fountain and peering into it to see this thing from different positions.

It was not until he was quite a good-sized animal that the pretence of killing his guardian was given up. As the gravity of age grew upon him, and those engaging pastimes of his childhood gave way before the development of inherent traits, these playful hunts became more rare and finally ceased. Both of us fully understood that this stalking business was nothing but fun. In fact, Gato never fully entered into the spirit of his part or displayed his powers to their greatest advantage, unless he was closely watched. Then, however, his acting was perfect. He got as far off as possible in the long, gallery-like room, fastened his glowing eyes upon the pretended victim, and from first to last showed how complete are the teachings of heredity, both in all that he did and avoided doing. Nothing that could favor his approach was neglected, no mistake was made. The furniture might be differently arranged with design, lights and shadows changed, new places of concealment, from which he could make his mimic attack, constructed; but the animal's tactics never failed to alter in accordance with these arrangements, and to be the best that circumstances admitted of. There is no doubt that he admired himself greatly, and, so far as it was possible to judge, commendation was very pleasing. He always expected to be complimented and caressed after darting from an ambush which had been reached with much precaution, and he reared up and rubbed his head against his friend, asking for praise as plainly as possible.

This account is not intended to convey any principles of zoöpsychology, but to record special facts relating to an animal whose family the author looks upon as exceptional in respect to their savagery, and who was himself, so far as the closest observation will warrant one in making a sweeping statement about a wild beast, not recognizably different in his characteristics from other members of the race he belonged to, or average individuals of allied species. "*Magnum hereditatis mysterium*" is a truth relating to the *process* of heredity alone; it has nothing to do with the fact that like produces like, or that traits, from the most generalized to those which are special, are undoubtedly transmitted. Here was a creature developed through immemorial generations into a typical state of body and mind. So far as the result is concerned, it does not make the least difference whether this end was attained in the manner pointed out by Darwin, or Galton, or Weismann. In Gato the whole personality, every faculty and feeling, the functional and structural peculiarities of all his tissues and tissue elements, were stamped with that impress which the entire life of his savage ancestry entailed. On what grounds can it be supposed that such perfectly superficial influences, as were brought to bear upon him while under restraint, produce any radical change? The alteration in demeanor manifested towards one person, and probably effected through that self-interest which, in its general aspect, is exhibited by all the higher animals, did not show that he had been, so to speak, inoculated with civilized sentiments. On the contrary, he gave a flat denial to that opinion every day, and was as essentially a puma, pure and simple, at the hour of his death, as if he had never seen a man.

It would, however, be a singular course of reasoning by which the inference that all pumas were the same was drawn from this statement. Besides the congenital variation that, to conceal our ignorance, we say is involved in the plasticity of life, every organism has certain acquired differences. Life is no more than a state dependent upon continuous adjustments, and it can never exist in an identical degree in separate beings, because neither the conditions themselves, nor the power to fit body or mind to circumstances, is ever the same in different individuals.

Evolved structures, functions, and qualities in groups, will be similar; for animals of all kinds must resemble their direct progenitors; but individuality is not extinguished, and as the race rises in capacity, or its members vary from an average, personal traits become salient, and those dissimilarities produced by alterations in the process by which existence is maintained, appear more prominently.

Almost the entire body of emotions which Gato possessed as a beast of prey, as well as his moral and intellectual traits, were beyond the reach of any modifications that could be made artificially. He was morose, cruel, treacherous, and blood-thirsty; but, it does not follow that he was absolutely so, or that, when compared with other pumas, these characteristics of his species were equally pronounced. Observation enables the writer to say that this animal was more intelligent, tractable, responsive, and reliable than any other beast of the same kind with which he ever was brought into close association.

A direct parallel between men, even barbarous men, and brutes will always fail. We do not know enough of the mental organization of either even to apply terms justly; and more than this, the difference between them in developmental states is so great that while the phenomena of both are of the same order, and the language used in describing one is applicable to the other, there are not close enough likenesses between them to make comparisons possible. Those who have attempted to frame psychological schemes, vitiated their work for the most part by a false method, or invalidated the conclusions arrived at, in consequence of preconceptions which biassed the temper in which evidence was examined. Dr. W. L. Lindsay ("Mind in the Lower Animals in Health and Disease") recognized the relationship between psychical manifestations wherever they took place, yet the influence, as in his case, of this, among many other hypotheses, was almost certain to make itself felt in the manner in which facts were regarded. On the other hand, Professor Prantl ("Reformgedanken zur Logik") excogitated a metaphysical system for beasts from the standpoint of an assumption that the chasm which separated them from humanity was impassable. He admits their resemblance in essential nature. He agrees that the dissimilarities which they exhibit are results of a difference in evolutionary degree, and then his whole argument goes to show that this is not the case, and that brute mind and human intellect are radically distinct in structure and function. As this analysis of the intelligence in mankind and inferior beings was made without reference to facts, it is not surprising that they should be traversed by these in all directions, and that almost everything which the Professor asserts to be impossible, is well known to naturalists as a matter of actual occurrence.

Gato himself set at naught many of his conclusions. He may not have exhibited either love, gratitude, sense of duty, or that spirit of self-sacrifice which dogs frequently, and other animals sometimes, display, and there was no opportunity for judging of his social instincts; but he certainly possessed the "time sense" that Prantl attributes exclusively to man. His account of periods and seasons was as accurate as possible; he measured intervals and knew when they came to an end. Whether the ability to count beyond three existed, it was impossible to determine. The three copper balls he used to play with were exactly alike, and if one was missing, its absence never failed to be noticed at once. If it occurred to him that it had been taken away intentionally, he got angry or sulky, as the case might be. During one part of his wardship, the periodical absences of his only friend put him out greatly, because, so far as actions revealed the creature's feelings, they interfered with his comfort. He became dangerous when grown, and occupied a room by himself, from which he was not removed while his guardian was gone. Under ordinary circumstances he was released for several hours every night, and when the time came, if there was any delay, he began to call upon his comrade to let him out, and grew fierce if not attended to. No one ever knew him to take any violent exercise in this apartment, but the gymnastic performances he went through outside were worth seeing. After being confined in solitude a couple of days, which was the length of time his friend generally remained absent, his eagerness to see him back became excessive, according to all reports. He was restless, savage, and sometimes refused to eat on the last evening. The servants said that long before they themselves heard the horse's tread, it might be known from Gato that his liberator was coming. But he never welcomed him as a dog or horse will do. He was full of exuberant vitality, endowed with an intelligent interest in the strange things around him, which he studied with continued interest, and inspired with an inherited passion for liberty. This always showed itself first. No sooner was the door opened than he darted out, intoxicated with being free, and it was not until nervous tension had been relieved by violent muscular motion, that he bethought himself of other matters.

To sit and watch a man take himself to pieces was pleasing but puzzling. It was evident that boots were part of the body, because his nose told him so. How could they be taken off, and why had these feet their claws behind? A sword and pistol did not perplex his mind, apparently, as much as the foot gear and spurs. The rapier he admired, like all bright objects, but the firearm excited distrust as being, perhaps, capable of going off spontaneously. He knew about revolvers, but placed no confidence in them whatever. Having presided over the strange process of taking off one skin and putting on another, inspected the articles of clothing removed, and assured himself that those assumed had really become part of the incomprehensible being who did these things, he was ready for his own toilet, which was confined to a gentle brushing of the head. This was expected, however, and was suggested if it did not come soon enough. Then he was ready to go to dinner, a pleasing interlude during which his manners were marked with the greatest elegance and discretion. It was not appetite that moved Gato — he had gratified that before; it was the performance itself.

Forks, for instance, those queer talons that were picked up and laid down, excited his curiosity. He examined them, he ate from one with propriety, their glitter attracted him, but he did not understand the rationale of such devices, and their use never failed to fix his attention. Moreover, on occasions when the amenities of social intercourse were in order, he was peaceable enough; not affable by any means, for he never noticed the attendants or appeared to be conscious of their presence. Smoking afforded this observant creature much satisfaction. Smoke itself, if puffed in his face, displeased him, but the preliminaries, striking a match, and the wreaths that floated away and vanished, all this he liked and pondered upon, as he did on certain pictures hung around, and everything that for reasons which can only be guessed at, excited wonder. Professor Prantl lays down the law that a beast cannot think logically; nevertheless, and apart from other facts which refute that decision, it was perfectly plain that Gato solved some problems implying this power. After a course of observations and experiments, it was discovered by him that shadows were not alive because they moved, and then these ceased to be pursued. Much study was requisite to arrive at a conclusion that the sunbeams reflected from a mirror were of the same inanimate nature. This was settled to his satisfaction only after great research. The creature saw this thing done time and again before convincing himself of the resemblance between those luminous shadows and the dark spectra which had formerly deceived him.

Gato grew graver with age, and abandoned many amusements in which he had at one time taken delight. It seemed to his guardian that there was a steady development of his intellect, which showed itself in everything he did. It would be too much to say that he was capable of thinking about his own thoughts, but who shall decide that he was not? With consciousness, memory, and a strong sense of personal identity; filled with innate tendencies, through the medium of which he interpreted external impressions; prone to contemplations that, as his eye and changing attitudes indicated, were not vague, apathetic dreams, no one can know that he did not revive mental states and meditate on centrically initiated ideas.

Personally, and so far as mere individual opinion unsupported by proof goes, the conviction in his friend's mind is that he did. Often, as with all cats, his brain was torpid. Unconscious cerebration, no doubt, went on, but only dim, transient images floated into the field of consciousness, and fragmentary, isolated, shadow-like pictures of outward things were presented to the "mind's eye." It was plain enough when he was in this semi-somnolent condition, and the difference between it and the active exercise of faculty upon something within himself, was unmistakable. He thought, but how, and about what? In his realm of that ideal world so little of which has been explored by man, subjective processes transpired such as we have no clue to and no measure for. The contents of mind, however, must be derivations from experience in a wild beast as much as in a human being. What he had observed, seen, felt, and remembered in that form which his own organization conferred, were manifested characteristically: that is to say, when vivid imaginations excited, or external sense-notices aroused him, the beast of prey awoke at once. The same most likely, or rather, most certainly, must have been true of all mental conditions, but while the animal remained impassive, the fact was indiscernible. When this savage warrior lay before his companion's arm-chair, and looked straight in his eyes with fixed intensity, calling to mind, perhaps, the things he knew about this man, it was natural that recollections of trainers' confidences, accounts given by travellers and hunters, one's own experiences, the many superstitions of civilized and savage peoples, should suggest ideas which had a tendency to color and distort observation upon the part of his *vis-à-vis*.

No one, however, who was not under the influence of a fixed prejudice, could have looked into Gato's unfaltering orbs and seen there any confirmation of the common belief that brutes such as he are only restrained by fear; or that they have an instinctive sense of reverence and awe in the presence of human beings. All the respect this one felt for his guardian he learned. Besides that, he had superstitions concerning him. In maturity his great size, and reports of the wisdom he had attained to, made the animal famous, so that many persons desired to see him — that is, through the grating at his door. But strangers found no favor with this misanthropist, and he disliked being stared at. Thus, after regarding such intruders with a stern countenance, and taking no notice of his friend under these degrading circumstances, he affected to be unconscious that anybody was there, or else deliberately turned his back upon the visitors. For a time it was supposed that this mark of contempt occurred accidentally. Gato could have had no conception of the significance of this act as it is understood in civilized society, but he did it for reasons of his own, and at length quite evidently on purpose.

As was said, curiosity, which is always indicative of

mental development, was an unusually prominent trait in his character. There were numbers of things to which he paid no attention, but when an act attracted his notice and was constantly performed, it appeared to require investigation, and he applied himself to the subject in a manner quite different from that superior air with which ordinary matters were regarded. Books amazed Gato. Nothing could be made out with regard to them by means of scent or sight: they were dead apparently, and not fit to eat. What was in them that never came out? Why should they be watched so closely? This question he never found any satisfactory answer to, and one might see that it often perplexed him. When he was little, reading made him jealous, and he put his paws on the page and invited his friend to play. This mysterious occupation lost its novelty in time, and the desire to romp passed away, but frequently in after days when he observed his companion turn towards the bookcase and get up, he escorted him to the shelves, scrutinized the way in which he looked for a volume, or turned over the leaves of several, and went back to see if anything was at last coming to light about this strange and constant occupation.

Gato resolutely refused to learn English. Why he preferred Spanish, no one knows, but he did, and would only respond to communications made in that tongue. Habit and association had much to do with this, no doubt, but there is reason to think that a distaste for our vernacular was one of the many prejudices which, in a measure, detracted from those qualities which embellished his character. His guardian discoursed to him at length; taught him to do and leave undone numerous things, but had to use the only idiom his pupil chose to acquire any knowledge of. If he were called in English, the perverse creature would not come. He stood and stared like an obstinate child. More than this, if he understood, as no doubt he sometimes did, and even wanted to do what was commanded, but could not, because he had made up his mind never to do anything unless spoken to properly, he got angry. There is no doubt in the writer's mind that this is a fact, and that the prejudice referred to existed. Force might have been resorted to, of course, but that would have had the effect of deforming his nature after every effort had been made to leave it to its natural expansion, except in so far as its tendencies were prevented from expressing themselves in homicidal acts.

Langworthy, "the lion-tamer," as the posters called him, used to say that feline beasts, after coming to know one, were infallible physiognomists, but that they had to learn a face before being able to understand its expressions; also that they only read the signs of anger and fear, and never looked for anything else, not caring about approval or kindness, because all the great cats were destitute of affection. Lions, tigers, leopards, and the rest, he believed, scrutinized the countenance chiefly to see if a man were afraid. If so, no assumed look could conceal the fact, and they instantly became dangerous. Privately he scouted the idea that there was any power to overawe animals in one person rather than another, and held that the sole difference between men in this respect depended upon quickness of observation, and especially upon fearlessness.

In the main this squares with what is known of comparative psychology, and of the *Felidæ* in particular. But like most sweeping assertions upon beasts or men, it is not wholly accurate. Many animals are exceedingly vain, nearly or possibly quite as much so as savage men, and vainer they could not be. Now this trait is inseparable from a desire for praise, and although it is no more necessary to feel any respect or affection for the persons who gratify this longing, than it is to love people because they are able to excite jealousy, creatures with such a disposition will always solicit approbation, and be pleased when it is accorded. Certainly this was the case with Gato, who was fond of display, and delighted in being noticed and admired; who did many things for the express purpose of being praised, and claimed commendation as plainly as if he had been able to speak.

The faces of brutes, similarly with those of human races which differ greatly in appearance from the observer, at first all look alike. But afterwards one begins to discriminate, and finally distinguishes differences between them, and changes in the same individual at different times. While Gato lay by the fountain listening to the wind murmur through the great tamarind boughs that shaded him, heard the water fall, saw the fleecy trade-wind clouds sail slowly overhead, and was evidently neither asleep nor lethargic, but keenly observant of every sight and sound, how easy would it have been to fit his reflections to the scene; "to opine probably and prettily," as Bacon expresses it, and provide the chained savage with poetic resignation, or indignant sorrow, to make him feel and think in forms as far from reality as the vapors that floated above him were far from being the substantial masses they seemed. Such writings, eloquent and interesting as they often are, do a positive disservice to science. Think, he did: that was to be seen in the eye that softened or grew stern; in its far-away or introverted expression, or quick scrutinizing glance; in the smoothed or corrugated brow, the quivering, contracted, or placid lips; in attitudes indefinably expressive, and variations of his ensemble that cannot be described.

How should human insight penetrate this underworld of the intellect? All things definite there were transmigrations of his own experiences under the stress of heredity. What was emotional, unformed, and yet operative, was the bequest of a wild and free ancestry that sent down their tendencies and traits, gave him his organization, and, with a certainty as inevitable as death, stamped everything that he could think or feel with their "own form and impress." His ideas were reproductions; his emotions rose into consciousness from unknown depths. The latter set him upon the verge of what his predecessors realized, vaguely revealed their past, prompted those unrecognizable half-memories that are born with every being, prepared him for possibilities from which captivity cut him off, stirred his heart, and made life and the earth all that they were or could be to him.

Varying phases of mind as outwardly evinced, manifested themselves clearly in Gato's behavior and in his changes of temper. Those serene meditations which had sway during beautiful days, and in the calm of tropical nights, bore little likeness to states of tension that sometimes possessed him when the storms of the rainy season set in. If an African lion is to be seen in his glory, he must be looked at by the lightning's glare. It is amid tempest and gloom that the full proportions of his nature come forth. So with this lion of another world. Many a time in the course of those nightly interviews which have been referred to, he roused himself from an intense contemplation of his companion, disturbed by thunder and the tumult without. Then while the wind blew unequally, roared through swaying branches, or mourned around the walls that shut him in, he quickened under the influence of over-tones in nature which human beings cannot hear. Storm and darkness wrought upon him as they will not do upon man. Beyond what was visible or audible, there was something that came from within himself; something that wove "the waste fantasies" of his dreams together, and gave character and purpose to ideation. He showed it in profoundly suggestive pantomime. But what "air-drawn" shapes were followed with those long, swift, soft yet heavy steps, on what his eyes were fixed, what feelings and fancies engrossed and transfigured him, gave that fierce energy, and led him in their train, are unknowable. They had no voice, but only with mute motions pointed backward to a past in which humanity shared no part, and which it cannot explore.

Those who have reared beasts of prey, must, it is probable, read works that describe the expression of their emotions with a certain dissatisfaction. Not for the reason that their authors lacked power, the art of observation, or scientific attainments, but merely because they themselves have seen and felt the influence of so much that is too evasive for definite detail. The grander passions may be painted; in virtue of the unstable equilibrium of nervous elements, and that comparatively imperfect system of connections existing between the centres, they are always explosive. But a world of complex, kaleidoscopic views interpose between fury and tranquility. Feeling cannot be continually intense, nor need it necessarily remain unexpressed because it is not violent. Only those emotions which are for the time absorbing have an unmistakable physiognomy, and these both brutes and higher beings feel but rarely. In attempting anything more than a suggestion of the impression produced by current feeling, the observer is liable to become constructive; to picture himself instead of the model, or to lose the subject in the midst of anatomical, physiological, and psychological details.

Unprovoked dislike, antipathy, permanent and constant in special directions, together with antithetical feelings, which are also said to be spontaneous, Gato possessed in abundance. He gave up trying to kill the Apostle John, but liked him no better than did those heathens who boiled the saint in oil. Whether on account of an animosity he had towards all men, or because in his own fashion he became superstitious about the statue, this much is certain, that if dragged up to it, he took offence. On the other hand, Gato made friends with a horse. Every morning when his groom let him out, Said trotted to the rear of the house, put his head over the half-door looking into the court-yard, and asked for a little wine and sugar with a gentle whinny. Sometimes Gato was chained to one of the buttresses of a tamarind and saw him. Often Said walked in on the stone floor and found him loose, as was customary while his guardian remained at home. At first, when actually confronted, the Arab showed a good deal of uneasiness. But the puma was then only half-grown, and upon being reassured, the horse concluded that it was all right, and paid no further attention to him. So this singular compact of neutrality was begun. On Said's part, it never became anything else. He suffered Gato when a mature and very large animal to walk around him, without any special recognition of his presence, and that was all. On the other hand, the latter respected, or admired, or had some kind of a friendly feeling towards the horse.

In order that he might not remain in that benighted state in which his forefathers lived among wretched Olmecs, Chichemecs, and Otomies whom the Aztecs captured to sacrifice to their war god, it was deemed proper to instruct him in the use and effects of fire-arms. He approved of cartridges as playthings, and watched them put into the cylinder, but did not think for some time that they were the things that went off and made a noise and flash. When he saw a ball strike, he used to leap at the scar and look for fragments scattered by the shot. Finally, by dint of seeing ammunition exploded, and snuffing empty arms, Gato got some inkling that there was a connection between a pistol he saw charged, and certain effects. Still it is very doubtful whether in his opinion a loaded revolver was dangerous, until experience convinced him that it would kill. In other words, he was taught that which unreclaimed wild beasts find out for themselves everywhere on the face of the earth.

What finished his education in this way, was an incident that very nearly proved disastrous to himself. One summer morning while he was fastened in the court-yard, and his guardian sat reading in his sala, a large rabid dog dashed into the room from the street, and without noticing the motionless figure in a chair, rushed out by an opposite door towards the puma, who lay under a tree. Instant aid was necessary to prevent the latter from being bitten; for although at that time he would have torn the dog to pieces, as he had already done in the case of two or three others, this would not have saved him. He witnessed the whole affair; saw the revolver, the aim and flash, heard the report, beheld the dog fall, struggle a moment, and die. Afterwards its body was dragged nearer to him, so he could feel assured that life was gone. Then for the first time did a realizing sense of the potency of this instrument enter into his mind. Subsequent to this occurrence, it was for a while only necessary to wear a pistol to keep Gato at a distance. Once in an unlucky hour his guardian told a servant to aim one at him by way of experiment, and nothing but the promptest and most determined interference saved the man. Charles Darwin ("Expression of the Emotions," etc.) says that the physiognomy of fear among cats is difficult to describe because it passes so quickly into that of rage. In this case the transition was instantaneous, and a fine fury it was.

The blare of cavalry trumpets, the roll of drums, and

clang of bells, attracted Gato's attention and made him restless, but he was not "moved by concourse of sweet sounds." They possessed no meaning, and did not cause him to think or feel. To sing to him was a waste of time, and he looked upon a guitar as something that made an insignificant noise. If the strings were roughly and unexpectedly vibrated, the effect resembled any other sudden interruption of meditation or slumber. He was startled, and apprehension instantly took the form of anger, and then passed quickly when he saw what had disturbed his repose. All physiologists will agree with Spencer that "the existing quantity of nerve force liberated at any moment which produces in some inscrutable way the state we call feeling, *must* expend itself in some direction, *must* generate an equivalent manifestation of force somewhere." The feeling excited, whatever it may be, will flow in accustomed channels, and manifest itself in what Darwin describes as "habitually associated movements." This law, and that governing antithetical manifestations, is founded in the physical and mental organization of all creatures, and its expressions vary with the differences obtaining among those of different kinds.

Gato and the members of every species belonging to his family are primarily avatars of force. They inherit as predominant traits those feelings and faculties, those physical specializations and particular aptitudes, which tend to make violence successful. When any nervous shock let loose his energy, it flowed from the centres where it was stored through the most permeable tracts; those which had been most frequently traversed in the history of the individual and his race; and as this process was necessarily accompanied by corresponding movements, when the strings of a guitar aroused him suddenly, Gato involuntarily assumed the attitudes and exhibited the temper of an excited beast of prey. If startled, teased, or menaced, if impatient, angry, or even pleased, however different may have been the passing feeling, however variously it was expressed, his character always overshadowed him, and gave an air to every outward act; not always in those set forms which Camper, Le Brun, Bell, and Darwin set forth, but unmistakably, and, of course, by the same means through which the typical representations of passion take place.

That sedateness and inertia which, in *Felidæ* especially, soon supervene upon the restlessness of kittenhood, showed themselves in Gato at an earlier period than usual. This was in a great degree attributable to his rapid and enormous growth. The energy which under ordinary physiological conditions would have remained free to manifest itself in movement, was expended in building his frame.

Many times on looking up and meeting Gato's gaze as he lay upon a rug contemplating his friend, the expression of those fiery eyes suggested stories of fascination — Arab legends, African and Hindu superstitions about the mesmeric power possessed by tigers and lions. A good deal has been written on this subject which is not much to the purpose. But no one has shown, or can show, that this influence is impossible, or, as it suggested itself to the author in the course of some experiments upon his puma, that susceptible subjects might not, as in cases reported by Charcot and others, hypnotize themselves. Having no way of getting at the relations subsisting between the centres of his brain with any certainty, it occurred to his guardian that a physiological approximation to their state might be made by means of this kind of an impression, and that it would reveal, to a certain extent at least, what is called by French writers the "solidarity" of that organ. The difficulty lay in the first necessary step, according to Heidenhain; namely, in arresting attention. Czermak's experiments at Leipzig were made upon creatures of a very different character from Gato. By all accounts, hypnotism is impossible except when attentiveness approaches to a wrapped degree of fixedness. The author tried to act upon his puma, but in vain. A bright object placed above him in front might or might not excite special curiosity. If his keeper held it, he looked at him, and probably wondered what new deviltry he was after then.

Often he grew uneasy, or disgusted perhaps, got up, and lay down with his head averted, or closed his lids. Sometimes he walked away, pretending not to notice his companion, though keenly observant of what he was doing all the while. But this eye-to-eye interview was quite as likely to bring the animal close, make him rub against his comrade, or present his head to be stroked. Whatever he did, however, was done of his own accord, and had no reference to the performances of his associate, or to the will-power exerted and wasted on such occasions.

It was easy to see when Gato was apathetic, and plain enough when he was intoxicated with what Willis and the old anatomists called "animal spirits." In the mean between these extremes lay the mystery. Who was to decide when the panther patted you gently with his sheathed paw, or put his head before the book, whether these solicitations to take notice of him had their root in a need for sympathy, or were signs of a desire to enjoy some pleasant sensation, such as being scratched or played with? One could only guess at this from the clue given by a knowledge of his character.

Much uncertainty exists with regard to the degree in which his æsthetic sense was developed. Whoever has shown pictures to children and savages, knows the great uncertainty attending their recognition of things which are familiar to them. The puma liked glaring colors and bright objects, yet while capable of identifying a large statue, the preference he exhibited for certain paintings depended most probably on their florid style. If his guardian read a work illustrated with engravings while he looked over his shoulder, they made no perceptible impression upon him. He admired gorgeous parrots that cursed him, and for a long time made hostile demonstrations towards a raven who was too wise not to let him alone. Some of the great hunters have thought that those strong predilections exhibited by tigers for certain beautiful localities which otherwise had nothing to recommend them to the choice of such inmates, were evidence of appreciation upon their part of this advantage.

That conclusion is, however, a very uncertain one, and most likely comes under the head of those observations that Czermak designates as "events viewed unequally";

that is to say, the facts are true, but the inference unwarranted. Gato had not much opportunity of studying nature. When, as happened several times during early life, he was taken into wild and solitary places, his attention concentrated itself upon living things. Beside those he seemed to care for nothing, except, perhaps, to be perverse. He climbed trees and would not come down, hid, and pretended not to hear when he was called. Once, improbable as it seems, he lost himself, and when all hope of recovering him appeared to be gone, here came the little wretch, in a very bad temper, nosing out his friend's trail, and convinced that he had been tormenting him, and done the whole thing on purpose.

It is time to close these memorabilia. Such facts as the records of his life contribute towards the ways of wild beasts, and illustrating their habits and character, have been now brought forward. A book might be written about his adventures and the traits he displayed, yet most of what was most interesting in his character lies on the border land of actual observation, and cannot be distinctly stated.

The manner in which Gato departed this life was worthy of himself, and may be taken as the last proof of his unchanged savagery of spirit. He had never come into actual conflict with a man, not because of unwillingness, but in consequence of the restraint imposed by confinement, bonds, or his guardian's presence. On the evening of his death he was fastened by the fountain; when, as it is said, for unhappily the writer was absent, a strange dog appeared, whom he sprang at, breaking his chain close to the collar, and killed. Afterwards he climbed a tree, and while the servants shut themselves up in their apartments, stretched himself out on a limb, and looked down upon the mangled remains of his victim. No doubt the ferocious feelings of his nature were all aroused, and unfortunately just at that time a man rode through the stone passage that in this country serves as a front door. Then the puma came down and flew at him, springing on to the croup of his horse, and wounding, though slightly, both it and its rider. The man being a nervous person, lost his head entirely, and not satisfied with making himself safe in a room whose door was opened to him, must needs fire out of the window with a carbine he found in the apartment. Some people become demented at the sight of their own blood, and this was one of them. He held straight, however, and the ball shattered the animal's right shoulder and passed backwards into his body. Gato had got between two great roots of the tree when his friend arrived, and that saved him from another shot. The creature was desperate, but too intelligent not to know that he who approached had no part in what he suffered. It was a mortal wound, but death promised to be delayed till that splendid frame was wasted by morbid processes and his life was gasped out in agony. This was not to be endured. The hand of affection did him the last good office, and he died instantly.

Pumas do not charge men in masses. Their victims are chosen among those creatures they find alone. Individuals have sometimes been assailed by more than one. Im Thurn asserts that the "Warracaba tigers" of Guiana, who hunt in families, are pumas. Two persons occasionally appear in authentic records as having been assaulted. Mostly, however, the incidents of any serious adventure of this kind are only known to a single individual, and whether they are ever recounted depends almost entirely upon the way in which the attack is made. A hunter taken by surprise would generally lose his life. This animal is not difficult to kill, and the facility with which it may be disposed of is another reason for disparaging its prowess among the class who most commonly encounter it.

A source of misunderstanding is also found in the special habits of this animal. Those of the *Felidæ* about which some more or less vague information is most generally diffused, do not climb. The puma is particularly given to doing so wherever forest lands are found within the range of its distribution. Quite as frequently as the Asiatic and African leopard, and more commonly than a jaguar, this beast resorts to trees when pursued. Its reasons for doing so cannot be doubted: it feels at home among the boughs; observation has taught the animal that none of those natural enemies it need avoid can follow. If dogs are on its track, it is well aware that, owing to their superior speed, they are certain to come up with it, and that in taking to the limbs above, its scent will be lost. For this habit but one reason has been commonly assigned; namely, that the puma is a poltroon.

In G. O. Shields' compilation of monographs upon "The Large Game of North America," he publishes some narratives that throw light upon the cougar's character. Revenge is not a very powerful or persistent passion in the

Felidæ, but cruelty is. Injuries are soon forgotten, and nobody ever knew a lion or tiger to act in this regard like an elephant. The feline beast never forgets, however, or becomes indifferent to the joy of torture. That is why it is fatal to fear it. The sight of this kind of suffering excites all their fell tendencies. Accidents with these animals are not results of abiding hate and premeditated vengeance, but very often of sudden impulse excited by the sight of apprehension. Deep, concentrated, persistent feeling is beyond the *Felidæ*. This is why Dio Cassius' story of Androcles and his lion is untrue; quite as much a romance of the affections as Balzac's "Passion du Désert." Gérard's touching account of his reunion with Hubert at the *Jardin des Plantes* fails, in his version of the animal's feelings, for the same reason — because it is impossible. No doubt the lion he had reared was glad to see him, but that is not what is conveyed. The picture presented is too like that drawn by Homer of the behavior of Ulysses' dog, when his "far-travelled" master came back, an unrecognized stranger, to Ithaca. No wild beast of the cat kind ever sat for that portrait.

Shields and others inform us that on several occasions "panthers" have been known to accompany women and children for some distance, and play with them, caper about their paths, and pull at their clothes, without doing further harm than was produced by fright. That these creatures act under the influence of playful moods is certain, but that a wild beast should come out of the woods and in pure lightness of heart invite a perfect stranger to romp, appears to be improbable.

Without pretending to decide upon what the mental or emotional state under such circumstances really was, both the natural character of these beasts, and certain well-known devices, not only of theirs, but of allied species, suggest another explanation. One of the most common means for defence resorted to by this family at large, is an assumption of anger, and the pretence of attack — they try to frighten intruders whom they suspect of an intention to do them harm. When a puma crouches and bares its teeth it is not always enraged, but very frequently does this for the reason that it is uneasy, or dislikes what you are doing and wishes to put an end to something disagreeable by terrifying the objectionable person. It might then happen that a cougar would, when startled by an accidental meeting of this kind, assume an offensive attitude with the intention of intimidating the person met. If it succeeded, apprehension might easily give place for a time to its propensity towards torture, and the beast would then behave much in the same manner, apart from actual violence, as if in the course of its pursuit of prey this had been overtaken. Such situations, however, present none of the conditions that tend towards permanence. In default of speedy rescue, the partially aroused tendencies of the puma would soon become fully awakened, and its impulses break out in acts of bloodshed.

Various references have been made to that part of the education of feline beasts by which they are taught not to kill their human associates. One may read a great deal without finding much information on this subject. Most all of the professional trainers whom the writer has exchanged ideas with on this point were of opinion that fear alone would prevent these creatures from becoming dangerous; and it is customary to proceed upon this principle. As soon, however, as any single rule is attempted to be fitted to all cases, it becomes plain that it will not apply. The personality of a cat is not to be compared with that of a man; nevertheless, if one is reared without taking this into account, it will be ruined. Such beings differ so greatly in disposition and temper, in capacity, and the power and willingness to learn, that to force them all alike into a mould, causes mental and moral deformity with the same certainty that a similar proceeding would cause distortion of their bodies if the means used were material restraints to physical development. The system of terrorism is based upon the false assumption that fear is the only feeling which will affect the *Felidæ* deeply and permanently, and that this can only be excited in one way; namely, by severity.

The intercourse of an average keeper with the animals he has in charge is in most instances of the most limited description. His observations, if he makes any, are more likely to relate to their behavior as either submission or otherwise, than to their general conduct towards himself, and usually, all he has to communicate possesses little interest except to the visiting public, who are easily satisfied, and ready to believe anything. A trainer or tamer, although often an interesting person in virtue of his experiences, is not always an instructive one. As a rule, all that he knows is confined to what has presented itself in the course of a few simple instructions. Experiments are rarely resorted to, both the knowledge of how to conduct them, and the attainments by which their results could be properly interpreted, being from the nature of the case most generally wanting.

A young savage of the cat kind will naturally bite and scratch when enraged, and the only means of discouraging such practices are those of punishment, and a clear demonstration that its hostile attempts are unavailing. No creature belonging to this class could comprehend the difference between right and wrong in an abstract form. But notwithstanding that what is bad in itself is hidden from them, things forbidden come to be quickly learned, and this *malum prohibitum* no doubt influences their minds in much the same way that, allowing for the inequalities, ceremonial observances and rites affect those of savages. The latter are largely occupied in performing and avoiding a number of actions because they expect personal advantages to accrue in one case, and condign vengeance to be visited upon malpractice in the other. They are superstitious, and so is the brute. Over and above the benefits or penalties these know of, there are others which they imagine but do not know.

To become even in a measure acquainted with pumas, one must take a reasonably good-natured and intelligent specimen in its infancy, and train it as consistently as if it were a child; make it feel the folly and futility of violence towards its tutor, impress it with the constant experience that its tricks and stratagems always fail before that friendly but invincible being who watches over its life and sees everything. Excite the animal's curiosity and wonder, show it the difference between yourself and others, be just and firm and calm. It will never be anything but a wild beast; but if this is done, it will be such an one as cannot otherwise be met with. Above all, if the interest of this occupation is not enough to affect the risk necessarily incurred, if such a pursuit cannot be followed without apprehension, give it up at once. A loose beast of prey is not a fit associate for a nervous man.

x

THE WOLF

THE wolf represents the typical form among *Canidæ*, and it possesses all the ordinary characters belonging to this group in their highest degree of development. There is but one family in the *Cynoidea*, that of the dogs, and all species of his group fall within the limits of a single genus. "*Canidæ* display likenesses in structure nearly as great as those which the cats exhibit," remarks W. N. Lockington ("Riverside Natural History"). Professor Huxley has broken up the aggregate into two groups, dog-like or Thoöid animals, and the *Alopecoids* — those which most resemble wolves. These are marked off from each other by peculiarities of the base of the skull and those parts developed around it. *Canis*, moreover, is a genus which, while it varies very greatly among its included forms, is physiologically so nearly identical that, as Lockington observes, "there is no proof that any species of this family is infertile with any other."

Wolves are among the wildest, wariest, and most widely removed from human association of all animals. The question whether all kinds — red, black, white, and gray — are of one species or many, may be dismissed at once. Nobody is able to say what specific characteristics really are. *Canis lupus* is one of the most widely distributed of

THE WOLF.

living forms. His range encircles the world within the arctic zone, and it extends southward into the tropics in America; wolves roam over nearly all Asia, and at one time they were found throughout Europe.

"The common wolf," says Lockington, "is the largest and fiercest animal of the group, and the only one that is dangerous to man." Its average length is about four feet six inches, it stands rather more than two feet high at the shoulder, and it is a little higher behind than before. These dimensions vary in geographical varieties; the French wolf being smaller than the German; the Scandinavian larger, heavier, and deeper in the shoulder than the Russian; while wolves on this continent are not so large as those of the Old World. All Asiatic forms north of the Altai Mountains are modifications of the common wolf of Europe, and the same is true of black wolves in the Pyrenees and highlands of France, Spain, Italy, and Russia, as well as of the white, lead-color, black, and dull-red varieties of America. As a rule, the wolf dwindles and degenerates within the tropics. *Canis pallipes*, the Indian form, approaches the jackal, according to Huxley, more closely than the members of any other climatic group, and as Professor Baird remarks, the coyote — *Canis latrans*, replaces the jackal in the New World.

Finally, the wolf, though a flesh-eater and beast of prey, possesses traits of structure which distinguish carnivora less highly specialized than *Felidæ*. Unlike the cats, its limbs are long and less united with the body; freer in their movements, and adapted to running rather than to the short, bounding rush and spring of the latter.

Wolves are very powerful animals in proportion to their size; active, hardy, with strong and formidably armed jaws. Their senses are all extremely well developed, their speed is great, and the tireless gallop of the wolf has given rise to stereotyped phrases and comparisons in many languages.

Leaving now the zoölogical relations of wolves, their habits, character, and capacity present themselves for consideration. At the commencement of such an inquiry we find sources of information upon some of these points which are valuable in themselves, and in their general tenor conclusive.

Cuvier ("Règne Animal") asserts that the wolf is "the most mischievous of all the carnivora of Europe," and it would have been possible to know this from the folk-lore of those countries alone. In mythology and minstrelsy, in fireside story and local legend, wolves stand foremost among wild beasts in nations of the Celtic and Teutonic stocks. Their fierce visages look out from all the darker superstitions of the Old World, and echoes of their unearthly cry linger in the saddest of its surviving expressions of dread, foreboding, and despair. Hans Sachs called them "the hunting dogs of the Lord"; but this is a conception restricted to a single religion, and nearly everywhere from Greece to Norway, the wolf has been an object of horror and hate, an incarnation of evil, the emblem, agent, or associate of those unseen beings under whose forms terror personified unknown and destructive forces.

All this is not meaningless; great masses of men do

not combine to give a "bad eminence" to anything that is insignificant. They do not often fear harmless objects, and they never do so when these are familiar. Cuvier says in his description of the wolf, that "its courage is not in proportion to its strength." But it is certain that packs once howled at night around Paris, and tore people to pieces in her streets; that they ravaged, and killed man and beast, in every part of Western Europe, made public highways unsafe, and put travellers by forest roads in constant peril of their lives. When the traditions and myths referred to were formed, things were much worse in this respect than in Cuvier's time, and we may be absolutely sure that these animals' reputation rests on a strong foundation of fact. It was not the accident of an idle fancy that pictured gaunt gray wolves, dripping with blood, that bore the spirits of death upon northern battle-fields. Geri and Freki, the wolves of Woden, battened on the fallen in Valhalla. On earth and on high, fantasy grouped its most tragic conceptions around "the dark gray beast" of early Sagas; and it was believed that chained in hell, the Fenris wolf awaited that day when the demons of the underworld should be loosed, and with the bursting of the vault of heaven, "the twilight of the gods" would come.

Very little good has ever been said about a wolf. But on the Western Continent there is an almost complete absence of evidence to show that imagination was affected by this creature in the same manner as was common among European nations.

Henry R. Schoolcraft ("Indian Tribes of North

America") remarks that "the turtle, the bear, and the wolf appear to have been primary and honored totems in most tribes. . . . They are believed to have more or less prominence in the genealogies of all who are organized upon the totemic principle." None knew wolves better than the aborigines of this country, and it is most improbable that beasts which so powerfully affected the thoughts and feelings of men in a similar social phase elsewhere, failed to conduct themselves similarly here. The cause for this striking difference is probably to be found in the peoples and not in the animals; more especially as every element was present in the situation where the former were placed, that would have fostered the growth of superstition. "The Indian dwelling or wigwam," says Schoolcraft, "is constantly among wild animals, . . . whether enchanted or unenchanted, spirits or real beings, he knows not. He chases them by day, and dreams of them by night. . . . A dream or a fact is equally potent in the Indian mind. He is intimate with the habits, motions, and characters of all animals, and feels himself peculiarly connected at all times with the animal creation. By the totemic system, he identifies his personal and tribal history and existence with theirs; he thinks himself the peculiar favorite of the Great Spirit, whenever they exist abundantly in his hunting-grounds, and when he dies, the figure of the quadruped, bird, or reptile which has guarded him through life, is put in hieroglyphics on his grave post."

This is not an exaggerated statement, and the fact is that the wolf was not only a tutelar of gentes and emblem of their confraternity, but also, as in case of the fabled founders of Rome, a protector of helpless innocence. In the cycle of legends and myths that gather around the culture-hero Hiawatha, we find the pretty tale of the "Wolf-brother." When the orphan child had been forsaken by all who were bound through natural affection to cherish it, wolves admitted the deserted little creature to their company, and gave the food that supported its life.

With southern tribes the coyote takes the place of the northern wolf; and how it happened that this "miserable little cur of an animal," as Colonel Dodge calls it (" Plains of the Great West "), became the guardian of anybody or anything, passes understanding, unless it be due to the fact that there is more cunning and rascality wrapped up in its skin than exists in that of any other creature whatever. Nevertheless, it is true that this jackal of the West undoubtedly occupies the position spoken of. Dr. Washington Mathews (" Gentile System of the Navajo Indians ") has shown that a coyote is the tutelar of at least three gentes in this great tribe, and Captain John G. Bourke (" Gentile Organization of the Apaches of Arizona ") traced this animal in the same capacity through several branches of the Tinneh family. He found coyote gentes in the Apache, Apache-Mojave, Maricopa tribes, and among the Pueblo Indians as well; at Zuñi, San Filipe, Santana, Zia, and other places. In his " Notes on the Apache Mythology," Captain Bourke gives a clue to the undeserved honors which this beast has received. His researches make it plain that these natives fully appreciated its astuteness. The coyote made a bet with the bear and won it; and by its means, also, men were provided with fire. There was nothing Prometheus-like in his conduct on this occasion; not a trace of the spirit which prompted the Titan. Far from it; he stole a brand the celestial squirrel dropped, and set fire to the world.

Like other wild beasts, the wolf has suffered at the hands of those who have described him. Men who, according to their own showing, had the most limited opportunities for learning anything about them have so often pronounced authoritatively upon the character of this race, and have so constantly confounded observation with inference, that closet zoölogists are now provided with a body of extemporaneous natural history in which the real animal has become as purely conventional as an Assyrian carving.

Perhaps the only accusation which has not been made against this much abused creature is that of stupidity. Nobody ever suspected a wolf of want of sense; although Buffon ("Histoire Naturelle") says, "*il devient ingénieux par besoin,*" as if he knew of other and more gifted animals who exerted their minds without any need for doing so.

The common representation which people make to themselves of wolves, and which they are most apt to see in pictures, is that of a pack. There is little doubt, however, that packs are accidental and temporary aggregates. They are not composed of family groups. Their members merely unite for an especial purpose, and disperse when this is at an end. Moreover, it is exceptional to find large numbers together in America under any circumstances. Wolves consort in pairs or small detached bands, and pack temporarily and rarely.

Captain James Forsyth ("Highlands of Central India"), speaking of *Canis pallipes*, an animal whose specific identity with the common form Sir Walter Elliot and Horsfield deny, while Blyth and Jerdon very properly insist upon it, remarks that it is a relatively small and slender beast with comparatively delicate teeth. He gives a narrative of his personal experience which is utterly subversive of many sweeping assertions which have been made upon the subject of their habits and temper.

In the provinces referred to, wolves are very numerous, and are "a plain-loving species." They "unite in parties of five or six to hunt," and so far as his observations go, more than these have not been seen together. "Most generally they are found singly or in couples." The domestic animals upon which these chiefly prey are dogs and goats. "They are the deadly foes of the former, and will stand outside of a village or travellers' camp, and howl until some inexperienced cur sallies forth to reply, when the lot of that cur will probably be to return no more. . . .

"The loss of human life from these hideous brutes has recently been ascertained to be so great, that a heavy reward is now offered for their destruction. Though not generally venturing beyond children . . . yet when confirmed in the habit of man-eating, they do not hesitate to attack, at an advantage, full-grown women, and even adult men. A good many instances occurred during the construction of the railway through the low jungles of Júbbulpúr, of laborers on the works being so attacked, and sometimes killed and eaten. The assault was commonly made by a pair of wolves, one of whom seized the victim by the neck from behind, preventing outcry, while the other, coming swiftly up, tore out the entrails in front. These confirmed man-eaters are described as having been exceedingly wary, and fully able to discriminate between a helpless victim and an armed man.

"In 1861, I was marching through a small village on the borders of the Damoh district, and accidentally heard that for months past a pair of wolves had carried off a child from the centre of the village, in broad daylight. No attempt whatever had been made to kill them, though their haunts were perfectly well known, and lay not a quarter of a mile from the town. A shapeless stone, representing the goddess Devi, under a neighboring tree, had been daubed with vermilion instead, and liberally propitiated with cocoanuts and rice. Their plan of attack was uniform and simple. The village stood on the slope of a hill, at the foot of which was the bed of a stream thickly fringed with grass and bushes. The main street, where children were always at play, ran down the slope of this hill, and while one of the wolves, that one which was smaller than the other, concealed itself among some low bushes between the village and the bottom of the declivity, the other would go round to the top, and, watching for an opportunity, would race down through the street, picking up a child by the way, and make off with it to the thick cover in the nálá. At first the people used to pursue, and sometimes made the marauder drop his prey; but finding, as they said, that in this case the companion wolf usually succeeded in carrying off another of their children in the confusion, while the first was so injured as to be beyond recovery, they ended, like impassive Hindus as they were, by just letting the wolves take away as many of their offspring as they wanted.

"A child of a few years of age had thus been carried off the morning of my arrival. It is scarcely credible that I could not at first get enough beaters to drive the cover where these atrocious brutes were gorging on their unholy meal. At last a few of those outcast helots, who act as village drudges in these parts, were induced to take sticks and accompany my horse keeper, with a hog spear, and my Sikh orderly, with his sword, through the belt of grass, while I posted myself, with a double rifle, behind a tree at the other end. In about five minutes the pair walked leisurely out into the open space within twenty yards of me. They were evidently mother and son; the latter about three parts grown, with a reddish-yellow, well-furred coat, and plump appearance; the mother, a lean and grizzled hag, with hideous pendant dugs, and slaver dropping from her jaws. I gave her the benefit of my first barrel, and she dropped with a shot through both shoulders. The whelp started off, but the second barrel stopped him also, with a bullet in the neck."

Whenever wolves hunt in numbers, it is that one part may lie in ambush, and the other drive the game, or because they design to assail enemies they are well aware a few could not overcome. These packs only hold together for a short time, and their formation depends upon the accidental presence of several separate bands in the same vicinity who are attracted by a common object, or follow each other's motions like carrion birds. This is what happens in the neighborhood of remote and isolated settlements in Northern Europe, when human beings are the game they pursue. Within Russian forests and those which lie near lonely villages in Sweden, Norway, and Swedish-Lapland, small packs form as darkness veils the weird, melancholy, desolate beauty of winter landscapes. They meet irregularly, with the vague, fierce feelings of an excited mob. The band is brought together by howlings, and it sweeps outward into the open on an indefinite quest. Woe betide the wolf who gives out during this wild gallop, or slips his shoulder on the frozen crust. Desperation may enable him to conceal the accident for a few strides, but discovery is certain, and he is instantly torn to pieces and devoured. If a fresh trail be found, the pack follows it. Human voices or the sound of sleigh-bells brings down the wolves like a storm-driven cloud. Men often go out with drags fastened to sledges, and as their purpose is simply to kill, and they are prepared, and do not venture too far from the villages, these hunters generally succeed in their undertaking. But not always; many a sleighing party of this kind has not returned, neither men nor horses. Many a belated wayfarer and party of travellers have never reached their journey's end. A fleet horse will for a time outrun wolves, even when by stealthy approaches they have almost closed around him, and this the author knows from experience; but it will not answer to go far, for in that case the fugitive will certainly be caught.

Turning now to the most celebrated, as well as the largest and fiercest member of this family, we find that the Scandinavian wolf is in many places increasing in numbers, despite the various means which are made use of for its destruction. L. Lloyd ("Scandinavian Adventures") ascribes this to immigration from Russia and Finland. However this may be, recent writers still echo the lamentations of Olaus Magnus, and of quaint old Bishop Pontappidan ("Natural History of Norway") to the effect that the country is overrun by them. Thus Von Grieff asserts that in many localities "the wolf taxes the peasant higher than the crown," and J. A. Strom expresses himself to much the same effect.

A wolf will eat any sort of flesh, irrespective of its kind or condition, and when pressed by hunger he consumes vegetable substances also. Pontappidan says that one was killed whose "stomach was filled with moss from the cliffs and birch tops." Humboldt states that famishing wolves swallow earth like the Otomac Indians on the Orinoco.

It is the common or gray wolf — the only one known in Scandinavia, although at one time Nilsson attempted to erect its black variety, *Canis lycaon*, into a species — which those authors referred to speak of when deploring this creature's destructiveness. Lloyd thinks that it cannot be extirpated from the mountain and forest regions of Sweden and Norway. The animal is prolific. A female, after ten weeks' gestation, brings forth from four to six, and even nine cubs. They are born in burrows, inherit great constitutional vigor, and are well tended upon the part of their parents. Whatever else may be denied the wolf, some praise for domestic virtues cannot in fairness be withheld from him. He hunts diligently and disinterestedly for the support of his mate and young, and when these (which are at first nearly black and look like foxes, except that they have not a white tip to their tails) are able to travel, both parents carefully supervise their education. Various diseases are prevalent among wolves, and many die of sickness; but if it be true that hydrophobia is unknown among those of Northwestern Europe, their exemption from a disorder which afflicts this species in all cold, and even temperate climates elsewhere, must be looked upon as an unexplained fact. During the rigorous and prolonged winters of high latitudes large numbers starve to death. Men shoot, trap, and poison them at every opportunity; they often kill one another, and when the ice breaks up in the greater inlets of the north Atlantic and Baltic, multitudes of wolves that have been hunting the young of seals upon their frozen surfaces perish.

Buffon seems to have furnished the wolf's character ready made for use by subsequent writers, since these appear to have done little more than copy or comment upon his text. "*Il est naturellement grossier et poltron,*" he says, "*mais il devient ingénieux par besoin, et hardi par nécessité; pressé par la famine, il brave le danger*" — that is, it will come out of the depths of forests, and attack domestic animals. "*Enfin, lorsque le besoin est extrême, il s'expose à tout, attaque les femmes et les enfans, se jette même quelquefois sur les hommes; devient furieux par ces excés, qui finissent ordinairement par la rage et la mort.*"

Now if one reads, not all, for that would be impossible,

but a great many accounts of actual observations upon wolves, and has at the same time some personal knowledge of these brutes, the foregoing will prove to be unsatisfactory. When special traits, and especially those of courage and enterprise, are examined in books, flat contradictions begin to appear. Colonel Dodge ("Plains of the Great West") maintains that the gray wolf of America is an arrant coward. Ross Cox ("Adventures on the Columbia River") asserts that he is "very large and daring." Nobody has ever denied that wolves are formidable creatures which can be dangerous if they choose; what their annalists have done is to proceed upon the assumption that they are exactly alike everywhere, and give the general disposition and character of an entire race from a few scattered specimens seen by themselves in some particular localities. Under any circumstances it would be useless to discuss the wolf's courage without having previously settled what courage in a wolf is, and how it displays itself. Principle and sentiment have nothing to do with it; appetite and passion are its sole incentives. To compare it, then, with that of some savage warrior in whom a certain standard of action always exists, is unallowable. Yet this is continually done, not openly and avowedly perhaps, but evidently in effect.

Audubon ("Quadrupeds of North America") saw wolf-traps in Kentucky. "Each pit was covered with a revolving platform of interlaced boughs and twigs, and attached to a cross-piece of timber that served for an axle. On this light platform, which was balanced by a heavy stick of wood fastened to the under side, a large piece of putrid venison was tied for a bait." Visiting one of these pits in the morning, with its constructor and his dogs, three wolves, "two black and one brindled," were found to have been caught. "They were lying flat on the earth, with their ears close down to their heads, and their eyes indicating fear more than anger." It is said by Felix Oswald, ("Zoölogical Sketches") that pitfalls always cow animals. At all events, in this case, the farmer, axe and knife in hand, descended and hamstrung them. Audubon stood above with a gun and the dogs, to whom these helpless creatures were thrown to be worried. None of the captives made any resistance worth mentioning because they were such cowards! If a lion of the Atlas, however, comes ramping down upon an Arab *douar*, leaps over the fence of a cattle-pen, and finds himself at the bottom of a trench, he meets death with the same resignation. But that is on account of the dignity of his character. No mortal knows what either animal thinks or feels, and, since there is no difference between their demeanors, it would be quite as easy to make the death scene of the wolf poetic, and probably fully as much in accordance with the truth.

What has been said of fortitude applies equally to other qualities. It seems reasonable to allow wolves some part in deciding what enterprises they shall undertake, which way an attack ought to be made, and whether the risk of any adventure is likely to overbalance its advantages. They are very well acquainted with the business which it falls to their lot in life to transact, and since the days have gone when Greek lycanthropes, German währ-wolves, and French loupgarous appeared among mankind, not anybody is able to put himself in this animal's place so completely as to appreciate those motives by which it is actuated.

Wolves differ with their geographical position, with the peoples that come in contact with them, and in virtue of individual peculiarities. What has been done by them anywhere, might undoubtedly occur again if the conditions remained unaltered. Dr. Henry Lansdell ("Russian Central Asia") knew of Tartars on the steppes who rode down the wolf and beat it to death with their heavy whips. He likewise learned that shepherds in the Caucasus protected their flocks by means of dogs. Yet his native attendants, as he reports with some surprise, actually allowed themselves to become alarmed at the threatened attack of a pack on the road from Kabakli to Petro-Alexandrovsk.

T. W. Atkinson's views ("Oriental and Western Siberia") were not so decided, and his experiences in these latitudes had been different. He saw plenty of wolves in the valley of the Ouba, and they had followed his party on the plains of Mongolia. Cossacks assured him ("Travels in the Region of the Amoor") that travellers upon the steppe were often devoured, and bands of these grim beasts frequently gathered about his camp by night. On one occasion while hunting he observed a fine maral — the large stag of high altitudes in the Ac-tan, Ale-tan, and Mus-tan regions — run into by three of these brutes. "The ravenous beasts were tearing the noble creature to pieces while yet breathing," when two *bear-coots* — black Tartar eagles — sailed over the spot, and one swooped. "The wolves caught sight of them in an instant . . . and stood on their defence. . . . In a few seconds the first *bearcoot* struck his prey; one talon was fixed on his back, the other on the upper part of his neck, completely securing the head, while he tore out the liver with his beak. The other eagle seized another wolf, and shortly both were as lifeless as the animal they had hunted."

This explorer, however, so far departed from the rule in such cases made and provided, that he did not immediately generalize the character of all the wolves in Asia from his observations of those two that permitted themselves to be killed by a pair of birds. On the contrary, when a pack followed his party in Mongolia, he was prepared to look upon it as a serious matter. They were in camp, the weather was mild, game abounded, and it was a beautiful night. "Before long we could hear their feet beat upon the ground as they galloped towards us. In a very short while the troop came up and gave a savage howl. The men now placed some dry bushes on the fire (which had been allowed to sink by the Kalmucks and Kalkas, lest its light should attract robbers), and blew it up into a bright flame which sent its red glare far beyond us, disclosing the wolves, their ears and tails erect, and their eyes flashing fire. At this instant I gave the signal, and our volley was poured in with deadly effect, for the horrible howling they set up showed what mischief had been done. We did not move to collect our game — that might be done in the morning. Our pieces were reloaded as fast as possible, for the Kalmucks warned us that the wolves would return. We could hear them snarling, and some of the wounded howling, but they were too far away to risk a shot. The fire was let down, and we remained perfectly quiet.

"We were not long left in ignorance of their intentions. Shortly there was a great commotion among our horses, and we discovered that the pack had divided and were stealing up to our animals on each side, between us and the water. The Kalkas and Kalmucks rushed up to our steeds, uttering loud shouts, and this drove the wolves back again. It was now necessary to guard the horses on three sides, as we could hear the savage brutes quite near. The men anticipated that they would make a rush, cause the animals to break away, and then hunt them down. A Cossack and Kalmuck turned to guard the approaches on each side, and I remained watching at the front. The fire was relighted and kept in a constant blaze by Kalkas adding small bushes, and this enabled us to see as well as hear our savage enemies. Presently I discovered their glaring eyeballs moving to and fro, nearer and nearer; then I could distinguish their grizzly forms pushing each other on. At this moment the rifles cracked to my right, and the fire sent up a bright blaze, which enabled me to make sure of one fellow as he turned his side towards me. I sent the second ball into the pack, and more than one must have been wounded from the howling that came from this direction. The other men had also fired, and I did not doubt with equal effect, for it was certain that they would not throw a shot away. In a few minutes the growling ceased, and all was still except the snorting of some of the horses. Both Kalkas and Kalmucks assured me that the wolves would make another attack, and said that no one must sleep on his post.

"To increase our difficulty, we now had but few bushes left, and none could be obtained near us; therefore it would only be by a most vigilant watch that we could now save our horses. The night, too, became very dark, and nothing could be seen at a short distance except towards the lake, where any dark object might be observed against the dim light that rested on the water. Sharp and keen eyes were peering out in every direction, but no wolf was seen, nor sound heard. The Kalkas said the wolves were waiting till all was still, and then they would make a dash at the horses.

"We had been watching a long time without the slightest movement, when two of the horses became uneasy, tugging at the thongs and snorting. The clouds rolled off, the stars came forth and reflected more light upon the lake. Presently howling was heard in the distance, and Tchuck-a-boi declared that another pack of wolves was coming. When they approached nearer, those that had been keeping guard over us so quietly began to growl, and let us know that they were not far away. As it was now deemed absolutely necessary to procure some bushes, four of my men crept quietly along the shore of the lake, two being armed, and in about ten minutes they returned, each of them having an armful of fuel. The embers were rekindled, and material placed on them, ready to be blown into a flame the moment it was needed. The sounds we heard in the distance had ceased for some time, when suddenly there was a great commotion. The other wolves had come up, and the growling and snarling became furious. How much I wished for light, in order to witness the battle that seemed likely to ensue. For a time there seemed to be individual combats; but there was no general engagement, and soon all became still as before. Again we waited, looking out for more than half an hour, when the horses began pulling and plunging violently; but we could see nothing. The men now blew up the embers, and in a few minutes the bushes burst into a blaze, and then I saw a group of eight or ten wolves within fifteen paces, and others beyond. In a moment I gave them the contents of both barrels, the others fired at the same instant, and the pack set up a frightful howl and scampered off." Atkinson found eight dead bodies next morning, and the bloody trails of many wounded that had gone off.

How would this party have fared if instead of warm weather, and the presence of a pack that merely desired to gratify their taste for horse flesh, and showed their willingness to brave fire and rifle-balls to this end, the steppe had been snowy and the animals starving? There seems to be no more doubt that a considerable detachment of Russian infantry was destroyed by wolves about fifty years ago in the passes of the Ural Mountains, than there is that the dragoon by whom Wellington sent his despatch after the battle of Albuera was eaten, together with his horse. "Daring as the wolf was in olden times," says Lloyd, speaking of that found in Scandinavia, "he has lost nothing of his audacity at the present day." In proof of which he collects from newspapers, parish registers, official reports, and the testimony of eye-witnesses, a statement of the ravages of wolves among domestic animals and human beings that almost equals those mediæval notices in which their evil deeds have been recorded from one end of Europe to the other. None of these, or rather, none the writer has met with, rival that recital given by James Grant ("The Wild Beast of Gevaudan"). French, Dutch, Belgian, and English journals, during 1765, were full of those events of which a brief abstract is inserted, and their prolonged occurrence finally came to be an affair of grave importance to the government of France.

In that year a beast, not identified as a wolf until after its death, created a reign of terror in the forest country of Provence and Languedoc, devouring eighty people about Gevaudan. "*Qui a dévoré plus que quatrevingt personnes dans le Gevaudan,*" says the official report. A drawing (from description) was sent to the Intendant of Alençon, and as this looked more like a hyena than anything else, it was given out that one of these brutes was at large. The province offered a thousand crowns for its head, the Archbishop ordered prayers for public preservation, and the commanding officer of the department scoured the country with light cavalry. These measures failed, and after a troop of the 10th dragoons had pursued it for six weeks through the mountainous parts of Languedoc, and though it was seen several times, had failed to come up with the animal, the reward was increased to ten thousand livres, and Louis XV. offered six thousand more. High masses innumerable were said, and cavalry, bands of game-keepers, and gentlemen with their servants, sought the monster in all directions. Hunters by thousands were in search of it for months, and in the meantime its howl was heard in village streets at night, children and women were killed in their farmyards, woodcutters lost their lives in forests, and men were dragged out of vehicles on the public roads by day.

At last the Sieur de la Chaumette, a famous wolf slayer, appeared upon the scene. His two brothers accompanied him, and they actually found and wounded the animal. The chase was taken up by him again, and he was joined by a party of hunters picked from the most expert foresters of fifty parishes. It was in vain, however, for they never viewed their quarry again. In September, 1765, the Sieur de Blanterne, in company with two associates, shot the wild beast of Gevaudan, which had ravaged a large region of Southern France for nearly a year. The carcass was sent to Paris, and proved to be that of an enormous wolf.

A creature capable of killing one man, is able, all things being equal, to kill a dozen or a hundred.

Wolves' ravages are at present confined to places from which we have no reports, and that is the reason why public opinion always places such occurrences in the past. In all essentials wolves are potentially the same as ever, but their relations to mankind differ according to geographical position. In one place they are harmless and timid, in another they are aggressive and dangerous. Throughout the Arctic regions of the earth, where one might imagine that privation would render them audacious, they generally avoid the presence of human beings and are not often seen. Franklin, Back, and Parry have little to say about them, and it is the same with many other travellers in their northern haunts. Bush, Kennan, Cotteau, Seabohn, Collins, Price, etc., have no information of any importance to give. Even Dr. Richardson, the naturalist, passes them by nearly unnoticed, and Rink ("Danish Greenland"), in his collection of the "Tales and Traditions of the Eskimo," is silent on this subject. All these authors, however, refer to other animals of the Arctic. Dr. Harris ("Navigantium atque Itinerantium Bibliotheca") finds places for the bear, musk-ox, fox, wolverene, in his immense repository of facts and impressions, but none for the wolf.

A somewhat comprehensive acquaintance with what has been said concerning this creature, disposes the writer to think, that the silence of explorers with regard to a beast that would naturally attract attention, is explained by Captain Ross ("Voyage to Baffin's Bay"). In his first expedition the wolf is not mentioned among those animals described in the "Fauna of the Arctic Highlands"; but in his narrative of the "Second Voyage" he says, "the perpetual hunting of the natives seems to prevent deer, together with those beasts of prey that follow on their traces, from remaining in their vicinity." Dr. John D. Godman ("American Natural History") contradicts Ross flatly, and asserts that "in the highest northern latitudes . . . wolves are very numerous and exceedingly audacious. They are generally to be found at no great distance from the huts of Esquimaux, and follow these people from place to place, being apparently much dependent upon them for food during the coldest season of the year." Godman does not say whether his information was got at first hand, or taken from others, but there is no doubt as to the fact that he is wrong. High latitudes do not furnish permanent habitats for game. Reindeer or caribou are not only migratory, but wander constantly; the latter being, as Charles C. Ward remarks, "a very Ishmaelite" in its habits. The same is true of other animals upon which wolves subsist, and the idea of their living in any numbers upon Eskimo leavings is amusing.

Milton and Cheadle ("The North-west Passage by Land") give much the same explanation as Captain Ross for the fact that wolves are so rarely seen in the far north. "Wild animals of any kind," they inform us, "are seldom viewed in the Hudson Bay territories, unless they are carefully tracked up. They are so constantly hunted, . . . and whenever they encounter man, are so invariably pursued, that they are ever on their guard, and escape without being seen." Forced to range widely because the character of this region involves constant change of place upon the part of their principal game, and made wary to the last degree by perpetual hostilities, it might well be that travellers found them absent from those regions they explored, and scarcely had an opportunity to observe such as were actually in their vicinity. Thus Parry ("Journal"), who was struck by their shyness, says, "it is very extraordinary that no man could succeed in killing or capturing one of these animals, though we were for months almost constantly endeavoring to do so."

Something, however, may depend upon local variety. Captain Koldewey ("German Arctic Expedition") tells us that "the peculiar — species, he calls it — of wolf met with in other arctic neighborhoods is not found in East Greenland; neither is the wolf-like dog now dying out from disease." Brown ("Fauna of Greenland") takes the same view, but whatever the facts may be, dogs and wolves have sometimes been known to treat each other very differently. Sir Edward Belcher ("The Last of the Arctic Voyages") saw a wolf, which he at first supposed from its appearance to be one of Sir John Franklin's surviving dogs, come up to his own team on the sledge journey of 1853. "It did not quarrel with them. . . . Its habits were certainly very peculiar; it cared not for us, and frequently approached so near that it might have been shot, but was not disposed to make friends." Even if the tameness of this animal had been due to starvation, that would not have accounted for the friendliness of Belcher's dogs. General A. W. Greely ("Three Years of Arctic Service") reports of his, that "whenever wolves were near they exhibited signs of uneasiness, if not of fear." Captain Ross noticed that his dogs at Boothia Felix "trembled and howled" whenever wolves approached them. It is well known, however, that in the arctic, as elsewhere, these animals interbreed. Godman gives the following: "*Scientia naturali multum versato et fide digno viro Sabina, se canem Terræ-novæ cum lupa coire frequenter vidis.*" Theodore Roosevelt and others speak of the same thing as coming under their personal cognizance.

In high latitudes of America and Asia the wolf's attitude

towards man is inconstant to a marked degree. Much difference is doubtless due to influences both general and local, permanent and temporary, which it is impossible to ascertain from any accounts. The packs C. A. Hall ("Arctic Researches") met with near "Frobishers' Farthest," and at J. K. Smith's Island, manifested none of that timidity which has been remarked upon as the consequence of constant persecution. On the contrary, "they were bold," says Hall, "approaching quite near, watching our movements, opening their mouths, snapping their teeth, and smacking their chops, as if already feasting on human flesh and blood." Similarly, "eleven big fellows crossed the path" of O. W. Wahl ("Land of the Czar") "one winter day, near Stavropol." They merely inspected the travellers and went on. Colonel N. Prejevalsky ("From Kulja across the Tian Shan to Lob-nor") saw but few wolves, and in his report upon the fauna of the Tarim valley, he remarks that they "are unfrequent, if not rare." During his expedition ("Mongolia"), however, the Tibetan wolf, *Lupus chanco*, the same animal he thinks that the Mongols of Kan-su call *tsobr*, but really the common species under one of its many changes of color, was found to be "savage and impudent." Captain William Gill ("The River of Golden Sand") saw "here and there" on the broken and undulating plains of Mongolia near the Chinese frontier, "small villages surrounded by a wall to protect them from the troops of wolves that in the desolate winter scour the barrens of San-Tai."

Nothing would be gained by multiplying references, which might easily be given *ad nauseam* without finding that there was any particular change in their tenor. Enough have been already presented to show how utterly valueless are those sweeping conclusions upon the character and habits of wolves, which we are too much accustomed to see. The widest generalization on this subject that can be made with any approach to certainty, is that these animals, over and above their specific traits, are what their situations and the experiences connected with ordinary and every-day life make them. It is a well-attested fact that the wolf may be domesticated, and instances of this kind are not uncommon. Audubon, for example, saw them drawing the small carts in which Assiniboin Indians brought their peltries into Fort Union. Samuel Hearne ("A Journey from Prince of Wales Fort in Hudson Bay, to the Northern Ocean") gives an account of certain things seen by himself, which seem to indicate that these animals occasionally bear like relations to savages with those which must have subsisted when they were first reclaimed. "Wolves," he says, "are very frequently met with in those countries west of Hudson's Bay, both on the barren grounds and among the woods; but they are not numerous. It is very uncommon to see more than three or four of them in a herd. . . . All the wolves in Hudson's Bay are very shy of the human race. . . . They are great enemies to the Indian dogs, and constantly kill and eat those that are heavy loaded and cannot keep up with the main body. . . . The females are much swifter than males, for which reason, the Indians, both northern and southern, are of opinion that they kill the greatest part of the game." This, however, cannot be the case, Hearne observes, because they live apart during winter, and do not associate till towards spring. "They always burrow under ground to bring forth their young; and it is natural to suppose that they are very fierce at those times; yet I have very frequently seen even the Indians go to their dens, take out the young ones and play with them. I never knew a northern Indian to hurt one of them ; on the contrary, they always carefully put them into the den again ; and I have sometimes seen them paint the faces of the young wolves with vermilion or red ochre."

This statement of the friendliness existing between man and these beasts is unique. James Morier in the mountains of Armenia, Persia, and Asia Minor, Douglas Freshfield in the Central Caucasus, Atkinson, Prejevalsky, and Gill in Northern Asia, Forsyth, Hunter, and Pollok in India and Indo-China, and a host of witnesses in Europe and America, have given evidence to their destructiveness and to the enmity with which they are regarded.

There never has been any question with respect to the wolf's intelligence. His sagacity and cunning are of the highest brute order ; and although, if one looks at a longitudinal section of his brain, it appears poorly developed, when compared with that of a dog, resembling, to use Lockington's simile, a pear with the small end forwards, the latter animal is probably not inferior to the former in natural faculty. "If we could subtract," says Professor Romanes ("Animal Intelligence"), "from the domestic dog all those influences arising from his prolonged companionship with man, and at the same time intensify the feelings of self-reliance, rapacity, etc., we should get the emotional character now presented by wolves and jackals." The former need to be wise in their generation, for it is but seldom that their "ways are ways of pleasantness," and their paths are never those of peace. Their gaunt frames and voracious appetites have become common colloquialisms, and each has to match his astuteness against all the devices for his destruction that human ingenuity can invent.

Lloyd describes the amenities and virtues that adorned the character of a wolf cub belonging to Madame Bedoire; how it guarded her premises, made friends with her dog, went walking with its mistress, played with her children, and howled when she did not caress it. The biography of this blessed infant was written by a lady; Lloyd merely inserts the account. It had to be shot when it was a year old. He himself had a young she-wolf whose most noticeable actions seemed to be connected with her endeavors to get pigs within reach of where she was chained. "When she saw a pig in the vicinity of her kennel, she, evidently with the intention of putting him off his guard, would throw herself on her side or back, roll, wag her tail most lovingly, and look like innocence personified"; but if, as occasionally happened, the pig's mind was impressed with these artless ebullitions of youthful joy, and it came near enough, the creature was done for. While Sir Edward Belcher's ship lay in winter quarters a wolf haunted her vicinity. He sat under her stern, he beguiled the dogs away, he drove off all the game. Then they tried to kill or capture him, but in vain. When pieces of meat were fixed at the muzzles of loaded muskets, he fired off the guns and ate the bait. Seated upon a hill, just out of range, this "charmed wolf," as the men called him, "narrowly watched the proceedings of those engaged in further schemes for his destruction, and exulted possibly in his superior wisdom." Belcher's sailors began to believe this animal to be one of the officers of Sir John Franklin's lost ship, the *Erebus*. Dr. Rae reports the case of a wolf that cut the string fastened to the trigger of a gun before taking the meat placed in front of it. And Audubon relates that wolves watch fishermen in the northern lakes, pull their lines up, and appropriate the catch. They gnaw through heavy timber into caches and undermine dead-falls. They uncover and spring steel traps, and are as difficult to beguile as the wolverene — it is impossible to say more. Captain Lyon's crew caught a wolf in a trap that pretended to be dead when the men who set it arrived. Wherever men carry firearms the wolf appreciates their effectiveness, and is perfectly well aware that his coat will not turn a rifle-ball. But while this exercises an obvious influence upon his general behavior, in most cases the ability to see the movements of his enemy seems to lessen his dread of what may happen. If several are together when fired at, they will scamper off; but it is very common to see them turn when they think themselves safe, and regard their adversary with strict attention.

Upon the whole, it is doubtful whether wolves have been much diminished in numbers anywhere, except in places where the country has become thickly settled. While these creatures have solitudes to fall back upon, they make use of those great advantages in the struggle for existence which they possess. Their speed, endurance, and hardihood, the number produced at a birth, and their exceeding sagacity, qualify this race to fight the battle of life, hard as it is in most instances, in a manner that but few animals of any kind can equal.

There are two reasons why, in the midst of fragmentary notices and romances innumerable, authentic annals of American frontier life are so meagre in their accounts of what these beasts have done. The first is that our earlier settlers were men such as they have encountered nowhere else, and the wolves were soon cowed. In the second place, perils threatened those living on the border, which were so much more imminent than any which ever became actual through the agency of wolves that these beasts came to be disregarded. Those depredations and murders which they really perpetrated were only perpetuated in tradition, and when survivals of this kind came to be recast by writers who, besides being unacquainted with all the facts, knew nothing about the animals themselves, they at once assumed a form that was stamped with all the incongruities of crude invention, and served only to conceal more effectually that portion of truth upon which these poor fictions were constructed.

It is probable that all, who, having really observed the character of those wolves that inhabit what were once the buffalo ranges of the Northwest, and then going southward made the acquaintance of that large, yellowish-red wolf called the *lobo*, in Mexico, will admit that there is much difference between them. In the Sierra Madre two wolves are commonly considered to be a match for a man armed as these people usually are, and unless the whole population have conspired together for the purpose of propagating falsehoods on this particular subject, it must be believed that the lobo is often guilty of manslaughter. It has not happened to the writer to be personally cognizant of the death of any victim of theirs, but riding westward one day through the forests of that mountainous country lying between Durango and the Pacific coast, in the interval between two divisions of a large train of *arrieros* separated from each other by a distance of several miles, a woman and two children, boy and girl, were met. Struck by the beauty of the little girl, and knowing the way to be unsafe, some conversation took place in which the mother made light of those dangers suggested, and declined, with a profusion of thanks, an offer to see the party safe to her sister's rancho in a neighboring valley. They had only a little distance to go along the ridge, she said, and would then soon descend to their place of destination. The wolves were like devils, it was true, but robbers were worse, and she had many times crossed there from her home without meeting with either. In short, — *muchissimas gracias Señor, y todos los santos, etc., etc. Adios!*

All of them were devoured a very short time after. Their clothes and bones were found scattered on the trail which they had not yet left before they were killed. The muleteers in rear who found these fragments collected and buried them, putting up the usual frail cross which is to be seen along this route, literally by scores.

This term *lobo* is indiscriminately applied in Spanish America to creatures that bear little resemblance to one another. The *guara* of Brazil is known under that name, an inoffensive, vegetable-eating animal, in every respect unlike the wolf in character and habits, and, according to Dr. Lund, specifically distinct from it in having the second and third vertebræ of its neck characteristically elongated. Emmanuel Liais, however ("Climats, Géologie, Faune du Brésil"), states the chief contrasts between those creatures in question succintly, as follows: "*Au point de vue du régime alimentaire, les deux espèces du genre Canis les plus éloignées sont le loup commun d'Europe, animal féroce et sanguinaire, et la plus carnivore de toutes les espèces du genre, et l'Aguara ou Guara du Brésil — Canis Jubatus de Demarest, appelé à Minas-Geraes très-improprement Lobo (nom portugais du Loup), et décrit par la plupart des ouvrages de mammologie comme le loup du Brésil. C'est cependent le moins carnivore de tous les chiens connus, et sa nourriture préférée consiste en substances végétales.*"

As has been said, the wolf does not reach its highest development in hot countries. Wolves may be dangerous and destructive within low latitudes, as is the case both in America and Asia, but it will be found that when this occurs their range is generally confined to elevated regions in those provinces. Major H. Bevan ("Thirty Years in India") states that "wolves are amongst the most noxious tenants of the jungles around Nagpore, and they annually destroy many children; but they do not commit such ravages as in northern India." The same is true of the "giant wolf," *Lupus Gigas*, that Townsend and other naturalists described as a distinct species; but this brute which has so evil a reputation in the highlands of Mexico, "the red Texan wolf," as Audubon calls it, does not extend in the United States to the northern prairies; it only exists as a variety of the common species in the lower Mississippi valley, and farther south.

Audubon remarks that this form of the common species has "the same sneaking, cowardly, yet ferocious disposition" as other wolves; nevertheless those anecdotes with which he intersperses his descriptions are certainly not calculated to foster the belief that his impression agrees with facts.

There are certain traits and habits belonging to wolves at large which may now be brought together. They are not by any means strictly nocturnal animals, but very commonly prowl by night, and in places where large packs assemble; most of what has with truth been said against them occurred under cover of darkness. By all accounts, it is amidst gloom and storm, while the *buran* rages over the arctic tundra, that troops of these fierce creatures do their worst among Yakut and Tungoo reindeer herds. Caribou are not herded, and have been but little observed by those who could give any information upon such a point as this. Everywhere, a wolf is destructive, fierce, wary and sagacious. Moreover, it will often become aggressive and audacious in the highest degree, when circumstances contribute to foster the development and facilitate the expression of its natural character. It is the typical wild beast of its family, and if it is not in many instances sanguinary and prone to take the offensive, there is a much better explanation for abstention from violence than that of natural cowardice. Wolves have far too much sense not to know what they can gain with least exposure to loss; and no beast of prey, that is sane, and not driven to desperation, ever proceeds upon any other principle than this. Given the existence of mind, those accidents by which mind is modified, and relative differences in degree among its qualities, must also be admitted. Comparative stupidity, evenness of temper, want of enterprise, tameness and timidity, undoubtedly distinguish wolf and wolf, as they do all carnivores. Still this would not account for the conventional wolf, or explain the anomaly of its imaginary character, or show why, or on what grounds, it is maintained that there should exist so great an incongruity in nature as an animal unadjusted mentally and yet adapted physically to a predatory life; that has at the same time the disposition of a tiger and the harmlessness of a lamb, that lives by violence, yet shrinks from every struggle, that maintains itself by the exercise of powers it must be fully conscious of possessing, and is constantly debarred from the results which it might attain through their exercise by causeless apprehension. This is very nearly what must be meant when a beast of prey is called a coward.

Wolves stalk their prey, ambush it, either alone or in collusion with others that drive the game, and they also run it down. The jaw is very powerful and formidably armed, and in proportion to its bulk this creature is exceedingly strong. A wolf, though structurally carnivorous, will eat anything — fish, flesh, or fowl, fresh or putrid, animal or vegetal. When he has gorged to the limit of his capacity, if anything remains it is commonly dragged to some place of concealment and buried. Then the brute lies down until the apathy induced by surfeit passes away. Wolves hunt both by sight and scent, by day and night. They will certainly interbreed with dogs, producing fertile offspring; and they may be domesticated. But as they grow older the characteristics germane to their savage natures assert themselves. It is said by Godman that "when kept in close confinement, and fed on vegetable matter, the common wolf becomes tame and harmless, . . . shy, restless, timid." If he had said it became *ill*, the statement would have been more conformable with fact. No such interruption of the normal course of life is possible without an impairment of health, both bodily and mental. Carnivorous animals are not to be turned into vegetarians at will, nor any creature's energies thwarted and cramped without distortion and atrophy.

Wolves no doubt can swim, but it is certain that a wolf seldom voluntarily takes to water in which he cannot wade. Audubon saw one swimming, and others have witnessed the like. Still all accounts represent these beasts as stopping short in pursuit on the bank of a stream. Naturalists say that the length of life in this species is twenty years, and it has been recorded also that they do not become gray with age. It looks like a purility to repeat what has been gravely reported more than once; namely, that when wolves have plenty to eat they get fat, become lazy, and are not so aggressive as under contrary conditions. On the other hand, nothing is more common than to find writers explaining every act of audacity as due to hunger. Most probably it is; they would hardly go hunting while in a state of repletion. But the question is, how these authorities find out the exact state of their dietaries, and can be certain that they must be starving before they will attack the wild Asiatic ox or American moose; also how much less food is required, to urge them on to assail a party of men.

In seasons of scarcity wolves of the northern plains prey upon prairie-dogs, ground-squirrels, hares, foxes, badgers, etc.; small creatures that offer no resistance, and which it is only difficult to catch. At the same time they hunt the large game of North America, and although, much to the disgust of a certain class of writers, the common wolf, which weighs about a hundred pounds, does not select a buffalo bull in the best fighting trim as an object for attack when a less formidable animal of this species can be found, or meet the moose, that often stands six feet at the withers, or indeed any creature that can kill him, in such a way as to give it the best opportunity for doing so, he often has to fight and frequently comes to grief. But they "give every human being a wide berth," says Roosevelt, and it would be strange indeed if they did not, since none are apt to be encountered who, according to the wolf's experience, are unprepared for offensive action, or who do not make it their business in those parts to destroy him. This fact has been completely realized by wolves of the plains, and it is for this reason that in these latitudes they have now become, what Colonel Dodge asserts that they are, "of all carnivorous animals of equal size and strength, the most harmless to beasts, and the least dangerous to man."

A wolf's structure is not by any means so well adapted to destructive purposes as that of the larger *Felidæ*. No species of the genus *Canis* has either the teeth, claws or muscles which belong to cats. A predatory animal may, and often does, make an error in judgment, but there is one thing that it never does, and that is, to attack deliberately knowing beforehand that it must fight fairly for victory, and that the issue is quite as likely to be fatal to itself as to its destined prey. A single wolf is not a match for those large animals it destroys; and when, in virtue of what Professor Romanes calls the "collective instinct," odds have been taken against them, they succumb before a combined assault.

Where parties of "wolfers," as they are called, pass the winter in placing poisoned meat in their way, and in localities in which they abound, destroy them for their skins by hundreds, wolves would need to be much less sagacious than they are, if what was noticed by Lord Milton and his companion was not true as a matter of course. "These animals," the account says, "are so wary and suspicious that they will not touch a bait lying exposed, or one that has been recently visited." John Mortimer Murphy ("Sporting Adventures in the Far West") had seven years' experience of the way in which wolves were shot, trapped, poisoned and coursed. The conclusion he came to from those observations which he relates so well, was that the wolf in such localities, "large, gaunt, and fierce as it looks, is one of the greatest cowards known." He omitted to mention — but Godman has rectified the oversight — that wolves carry their natural cowardice to such an extent, and are so exceedingly dubious concerning what man may do, that a few pinches of powder scattered about dead game, or an article of clothing left near it, — in short, any evidence of the presence of a human being will prevent them from approaching it.

There are several ways of writing natural history, and this is one of them. It would seem, nevertheless, that if a plan could be adopted for looking upon the general organization of wild beasts as in a great measure determining their characters, and for considering, if possible, anomalous traits as most probably intimately connected with peculiarities in their situation, we might no longer feel confounded at finding that sentient creatures are not the same under dissimilar circumstances. If brutes could be considered to have some knowledge of themselves, to act like brutes and to feel like them, without reference to any human opinions whatever, forthcoming literature of this kind would be benefited.

In those parts of the world where the wolf comes in contact with people not well prepared to receive him, his attitude towards mankind is aggressive. In Eastern Europe, for example, Austria, Hungary, Bohemia, and through the Danubian states generally, wolves occupy quite a distinguished position for dangerousness, and the inhabitants regard them with any other feeling than that of contempt. Captain Spencer ("Turkey, Russia, the Black Sea, and Circassia"), while passing through that vast forest which separates the more settled tracts of

Moldavia from the Buckowina, was besieged in a half-ruined chalet with his companions, and the pack continued their attack all night, and lost heavily.

The coyote, — *Canis latrans,* — that thieving creature which is often found intermingled with the gray and other coated wolves on the great plains of North America, has been by some writers — Colonel Dodge, for example — discriminated from the prairie wolf as a separate species. Those differences which exist between them, however, have little classificatory value. Contrasts in size, dissimilarities in color, marking, and the growth of hair, are all seen in the common wolf, of which this is "a distinct but allied species," with northern and southern varieties.

"There is," says Schoolcraft, "something doleful as well as terrific in the howling of wolves." When people speak of the jackal's howl, they commonly call it "unearthly," but a coyote's voice is much more singularly diabolical, and his intonations are so hideously suggestive of all that is wierd and devilish, that it stands by itself among natural sounds, and cannot be compared with the outcry of any other creature. Murphy describes it as follows: "The voice seems to be a combination of the long howl of the wolf and the yelp of the fox; but so distinctly marked is it from either, that, once heard, it is never forgotten. The coyote has the strange peculiarity of making the utterance of one sound like that of many; and should two or three try their larynxes at the same time, persons would fancy that a large pack was giving tongue in chorus. The cry appears to be divided into two parts. It first begins with a deep, long howl, then runs rapidly up into a series of barks, and terminates in a high scream, issued in prolonged jerks." According to conventional opinions, elephants among wild animals, and dogs among those that have been domesticated, occupy the highest places in order of intelligence. The author does not believe this to be the case with respect to the first named species, and so far as pure intellect goes "Die reinen Vernunft," no dog can probably surpass *Canis latrans*. Professor Huxley also reports that he can find no essential difference between their skulls. While these animals may be equal, however, in absolute capacity, the coyote, considered according to civilized standards of manners, is the kind of creature that if any dog were to take after, he would be incontinently shot or hanged.

His idea of good conduct is to get what he can honestly procure when driven to straightforward courses, but by preference to steal it, as being less troublesome. He is astute beyond comparison in nefarious practices, and has sense enough to howl with derision (as he sometimes seems to do) if it could be explained to him that mankind were capable of judging his behavior according to any other rule of life than his own. *Homo sapiens*, in a highly evolved state, is imbued with the truly noble idea that he is the centre of creation, and that all living things are admirable in proportion as they approach himself. He calls the coyote a "miserable cur," "a barking thief," and says sarcastically that the brute has kleptomania. Savage man, on the contrary, esteems him greatly. The two are much alike in many respects. We have already seen that this little wolf has been adopted as the tutelar of gentes among Pueblo Indians, and southern tribes of the Tinneh stock, and its prominence is scarcely less with those of the northwest coast of America. They honor the coyote; their myths and folk-lore record its good qualities and wisdom. To them it is the incarnation of a deity or a demon (these are nearly the same), and it is never killed, for fear that ill luck might be sent by the spirit of which this animal is the representative.

Under these happy auspices coyotes hang around native encampments and villages, interbreed with Indian dogs, grow fat on salmon cast upon river banks in the spawning season, hunt all that smaller game which their more powerful relations resort to for supplies only when hard pressed, and omit to take advantage of no opportunity to gain possession of provisions which are not theirs. The opinion they have of the human race is that it exists for their advantage, and mankind, further than it contributes to their support, is an object of indifference to them.

More to the south, and in the vicinity of white settlers, the coyote is oppressed and persecuted; subjected to like usage with that which the common wolf receives. This state of things is of course accompanied by changes in character that are not less marked than in the wolf's case. It becomes nocturnal in habit, flies from the face of man, and is one of the most wary, timid, and suspicious of animals. At the same time its cunning grows greater as the necessity for self-preservation becomes more pressing, and in the same measure in which it is pursued does its capacity for evasion enlarge. Speed, endurance, wind, and invention, all develop themselves. Unlike wolves, whose homes and breeding-places are commonly in caves or clefts of rock, beneath trees or within any natural recess, coyotes dig burrows in the open, and are seldom or never inmates of forests.

As the species approaches its southern limit, the average size decreases and its color changes. In Mexico, where they are seldom molested, these brutes prowl a good deal during the day; they pack likewise more commonly than further north, and if smaller, are also bolder and less upon their guard.

In Algeria or Oran an Arab knew when the lion was coming by the jackal's cry; Brazilian Indians tell one that they can trace a jaguar's way at night through the barking of foxes, and it is said by shikáris in India that a prowling tiger's path may be known by a peculiar howl which his frequent attendant — the kind of jackal called *Kole baloo* — utters on such occasions. The coyote also gives warning of the approach of foes that are oftentimes more dangerous than either lions or tigers. But it is by its silence that danger is announced. In a position where hostile Indians were to be expected at any time, when the coyote ceased its cries, it was an ominous thing, and frontiersmen looked out for the appearance of a war party. Everybody who has been much on the border is probably acquainted with this very general belief, and it may perhaps be founded in fact; but this much is certain, that these creatures do not always become quiet when Indians are about, for the author has more then once heard them howl — coyotes, not savages who were imitating them —

when it was known for certain that Indians were near, and when the fact of their presence was soon proved.

Coursing coyotes is a favorite sport with many persons in the West, and while the weather is cool and dry they often make good runs; otherwise, the game soon succumbs to heat, or to a serious impediment in the way of escape— its own tail. This is carried low, and despite his long hind legs and powerful quarters, the brush gathers so much mud in deep ground as seriously to embarrass flight.

In those localities where this race exhibits indications of much timidity, it will be found that every destructive device of man's ingenuity is practised against it; even to taking advantage of a harmless weakness for assafœtida in the matter of preparing poisoned baits. All this makes certain associations of ideas inevitable, and special impressions upon his mind things of course. At the same time, no mortal knows precisely what these are.

Where no such experiences of human malice and duplicity color the coyote's character, its conduct is quite different. Under those circumstances it does not fly from imaginary perils. Even when fired at it shows no unseemly haste to leave; but if the shot be repeated, then the hint is always taken, and it vanishes. Most persons who have become personally acquainted with them must have had occasion to observe that where they have been subjected to the worst that man can do, their dexterity in the way of robbery is not more striking than the audacity by which it is accompanied. It seems difficult to reconcile the idea of any instinctive fear of man with the conduct of an animal that will steal through a line of sentinels into a military encampment, and carry off food from beside watch-fires. They do this; they do everything that requires enterprise, judgment, and skill, and this to an extent that, in the mind of an unprejudiced savage, has gained them a place among his gods.

Once the writer saw as much of the temper of coyotes in their natural state towards man as it is possible for anybody to see at one time. It befell that he was badly hurt in front of General Treveño's cavalry brigade, then holding the line of the Rio Caña Dulce. When consciousness returned, horse and arms were gone, and the bushes around swarmed with these wolves. There may not, however, have been so many as there appeared to be, for the animals moved in and out of cover constantly, and the same one was probably seen several times. The thirst that always follows hemorrhage, and the heat of the sun, were distressing, neither was it pleasant to be an object of so much attention to a troop like this, while almost completely disabled. An overhanging bank lay near, and was reached with great difficulty. Here one could lean up against the side and contemplate them from a shady place. They behaved very curiously, and if the attendant circumstances had been at all conducive to mirth, their spiteful antics, the pretences of attack they made, and the absurd way in which some of them assumed an air of boldness, and apparently sought to inspire their companions with resolution, would no doubt have been amusing. It was abundantly shown that these creatures looked upon the inert and blood-soaked individual before them as a prey, and were consequently in a high state of excitement. Their eyes sparkled and the long hair around their necks bristled; they made short runs at and around the position, they pushed each other, and howled in every cadence of their infernal voices; also some individuals showed the rest how the thing ought to be done. A rush would have been at once fatal, but it was not made. Nevertheless, they grew bolder, and when relief arrived, had for the most part gathered around in the open. What would have happened when night came, or whether anything, the writer does not pretend to say.

THE GRIZZLY BEAR

BEARS are included by zoölogists in that order whose typical forms are, besides themselves, the dog, cat, and seal, and they belong to the higher of those sub-orders into which this group of carnivora has been divided. *Ursidæ* hold a middle place among bear-like beasts, and although their generic history is not so complete as that of others, Dr. Lund's discoveries in Brazilian bone-caves brought to light a fossil form that Wallace regards as representative of an existing American species. Their palæontological record carries them far back among the fauna of earlier geological periods, and connects the sub-ordinal section which contains existing arctoids with insect-eating and pouched vertebrates on one side, and on the other, with the precursors of monkeys, apes, and men.

In their most general structural traits bears possess the characteristic features of all carnivores — their abbreviated digestive tract, developed muscular systems and sense organs, and highly specialized teeth. At the same time this genus is considerably modified, and on that account bears were placed among *Fissipedia*, which are practically omnivorous. Finally, *Ursidæ* are plantigrades with muscles fused in plates, and so exhibit the ungainliness, the awkward and comparatively slow and restricted movements peculiar to the genus.

Geographically they are nearly cosmopolitan. Their species, although not numerous, inhabit arctic and tropical regions, and live in the lowlands of Europe, Asia, and America, as well as among the mountains of both continents.

The grizzly bear is confined to the New World, and there is distributed from about 68° north to the southern border of the United States, chiefly in the main chain of the Rocky Mountains and on their eastern and western slopes, but also among the ranges between these and the Pacific. It has been called by many names. Lewis and Clark, who may be said to have discovered this animal, speak of it indifferently as the white and brown bear. Cuvier said he was not satisfied that any specific distinction existed between the latter and our grizzly, which has also been identified with Sir John Richardson's "barren-ground" species of the Atlantic area. Audubon supposes *Ursus horribilis* to have formerly inhabited this province, but the only basis for such an opinion is found in his interpretation of some Algonkin traditions. The present title—horrible, frightful, or terrible bear—is a translation into Latin of George Ord's name *grisly*, given in 1815. As it is commonly written, however, its significance is lost, the reference being to color instead of character. Dr. Elliott Coues and others have remarked upon this discrepancy, but it is now too late to make a change. The naturalist Say ("Long's Expedition to the Rocky Mountains") first described this species, although its physical features are well given by Captains Lewis and Clark, and it was mentioned before their time. Since then the animal's dimensions have been often and also differently determined. Lockwood ("Riverside Natural History") very properly gives no ultimate decision. Lord Dunraven ("The Great Divide") speaks of having shot "a middling-sized beast weighing about eight hundred pounds." Richard Harlan ("Fauna Americana") says that the animal's "total length is 8 feet 7 inches and 6 lines; its greatest circumference 5 feet 10 inches; the circumference of its neck 3 feet 11 inches, and the length of its claws 4 inches 5 lines." Captain Lewis measured tracks "eleven inches long and seven and a half wide, *exclusive* of the claws," which are reported by different observers to be of all lengths between four and seven inches; and the truth is that no one has been in a position to pronounce definitely on a single point respecting this animal's weight and size. It is the largest and most powerful beast of prey in the world. So much may be said confidently, but beyond that data for positive statements are not extant.

With regard to the grizzly bear's habits, they are variable, like the color of his coat, which may at one time and place justify the name he bears, and at another be almost black. *Ursus horribilis* preys upon all the large game of North America; he is, as H. W. Elliott ("Our Arctic Province") observes, "a most expert fisherman," and appears to be equally partial to wild fruits and carrion. These brutes consume large quantities of mast, they dig up the *pomme blanche* and other tubers and roots, and it is said that their relatives of the black species are sometimes devoured. Nothing edible comes amiss to a grizzly, from the larvæ of insects to spoiled salmon, or from buffalo-berries to the animal itself. But it must be admitted that accurate information is wanting upon many particulars connected with his way of life. Hibernation, for example, which is a trait varying greatly in its completeness among species of different genera, appears to be absent in this case. These animals go about both by day and night, in cold weather as much as in warm. There are perfectly reliable accounts of their having been encountered at all seasons, and in situations which were peculiarly favorable for going into winter quarters if the animal had desired to do so.

Again, the grizzly's exploits as a hunter are involved in much obscurity. It does not require great skill for him to catch buffalo, or supply himself with beef on a cattle range. The *Bovidæ* in general are not particularly intelligent, and no doubt an ambuscade which might be successful with them is managed without much difficulty. With deer, however, it is not the same. Caribou and elk, the black and white tailed *Cervidæ*, are not to be had by any man without a previous acquisition of considerable knowledge, without the power to put this in practice according to varying circumstances, and without great practical dexterity in several directions. Bears are not exempt from the requirements pointed out. All that is true of instinct restricts itself in every instance of efficiency to the fact that transmitted faculty makes acquisition rapid and promotes the passage of deliberate into automatic action. Apart from the advantages he possesses in this way, a grizzly bear needs to learn in the same way as a man. There are occasions constantly occurring in which mind must be exercised in a manner such as experience has not prepared him to meet, and where the animal acts well or ill, successfully or unsuccessfully, according to his individual capacity.

John D. Godman ("American Natural History") calls it "savage and solitary." All the more powerful beasts of prey might be similarly characterized. The influence of organization, inherited tendencies, and their daily life, indispose creatures of this kind towards association. Moreover, they are most generally rivals in their usual habitats, both as hunters and as suitors during the pairing season. We have no accounts, like those given of lions and tigers, to show how males behave toward each other under the antagonisms implied in contact, but everything points towards conflict. Still, as there are conditions which bring the former together in certain localities, so grizzlies sometimes congregate. Möllhausen ("Diary of a Journey from the Mississippi to the Pacific") reports that at Mount Sitgreaves, and in its surrounding eminences, their dens were so numerous that Leroux (a famous guide and hunter of those days) had never seen the same "numbers living together in so small a space." They had all gone when Möllhausen's party was there, owing to the freezing of waters in that vicinity. Those places where they had tried to break the ice were often found, and many trails well marked in snow showed that the bears had "made their journey to the south in troops of eight or more," each detachment going in single file.

Nevertheless, "Old Ephraim," as mountain men call him, having inspired all who ever penetrated into his haunts with a wholesome respect, has naturally been exposed to misconstructions. His character is frequently represented as more fierce and morose than it really is. Writers say of him that he will not tolerate the presence of a black bear, or the variety of this species, according to Baird, the "cinnamon," in his neighborhood. They tell how their boundaries are sharply defined, and remark that occasionally small numbers of these less formidable members of the family live as enclaves within the grizzlies' territories, but are rigorously confined to their own limits.

This is one of those wholesale statements with which descriptive zoölogy is full. No doubt there are plenty of grizzly bears that would kill any poaching relative of theirs unlucky enough to encounter them. As a general fact in natural history, however, the theory of the separateness of distribution among American *Ursidæ* will not stand. Many direct observations show it to be otherwise, and Schwatka ("Along Alaska's Great River") is fully supported in saying that he doubts the truth of this statement from his own experience. On Cone Hill River he saw "four or five black and brown bears in an open or untimbered space of about an acre or two."

There are spots in India appropriately called "tigerish." Any one who knows the beast's ways would naturally look for it in these sites. But it is very doubtful if the physical features of localities have much to do with selection by this species, apart from the fact that when he feels himself to be in danger, a grizzly gets into the most inaccessible position possible. He loves cover under all circumstances, although it is not uncommon in secluded situations to find these animals far out in open country;

but timber and brush seem to be more or less accidental accessories so far as his preference is concerned. The animal needs a constant supply of water, and if this can be had, broken and intricate ravine systems suit it as well as thickets or forest land. Its partiality for swamps depends upon their productions, and the fact that game is apt to be found in them. Independently of special considerations of any kind, the propensity to conceal itself is a natural and necessary outgrowth of the habits and character of all predatory creatures. They do so universally, and a grizzly, like the rest, much prefers a windrow, precipitous arroyo, or brake, to any plain whatever which is not overgrown in some way.

Grizzly bears do not climb trees. They are said to shake them in order to procure fruit, and also for the purpose of dislodging men who have taken refuge among their branches; in general, however, the animal sits up and claws down the boughs within reach.

Probably that conventional expression, the "bear hug," has no significance anywhere. Some bears hug tree stems in ascending trunks adapted to their embrace, but Asiatic species of all kinds simply sink their claws into the bark of boles they would be utterly unable to gain any hold upon otherwise, and climb like cats. This arctoid is too heavy for that; he is over-sized, in fact, like the greater *Felidæ*, for any arboreal gymnastics. The writer can find no reliable evidence to show that this or any other bear attempts to inflict injury by straining the body of an enemy within its arms. A grizzly will grasp and hold a man or beast while biting, or striking with the claws of its hind feet, and blows from its forearm are delivered as frequently and not less effectually than is customary with the lion, but beyond teeth, talons, and concussion, no authentic mention is made of modes by which its victims are put to death.

All young vertebrates are playful in youth, and if taken early enough, some would be found even in species commonly regarded as untamable, that for a time at least might be domesticated. Among *Ursidæ* untrustworthiness is the rule. They are quite intelligent, capable of being taught, and competent to understand the necessity for being peaceable. Yet if one judges from reports they are more unreliable than the cats. Relatively these animals are not so highly endowed, and this fact, coupled with inherent ferocity, and an organization by which passion is made explosive, accounts for the character they bear. Cubs of *Ursus horribilis* grow savage very soon. Lockwood and others regard the species as incapable of being completely tamed. As far as that goes, however, the same is true of every wild beast able to do harm. These animals are kept under the same conditions as other show creatures, and seem to be in much the same state. It is nevertheless probable that either from a greater degree of insensibility or less mental capacity, they always remain more dangerous than most *feræ*. This brute has nothing of the phlegm about him that his appearance suggests. He is morose, surly, and rough at all times, and even more liable to sudden and violent fits of rage than a tiger.

Either, as seems likely from what we know of the animals in question, on account of the fact that those who have had an opportunity to observe them were exclusively occupied with describing their destructiveness, or because grizzlies have few of those traits that make many species interesting, their records are very barren indeed. A solitary being like this could not possess the engaging qualities Espinas ("Sociétés Animales") and Beccari describe among those that live in association; but other creatures are so placed without losing all attractiveness. It does not take long to tell the little that is certain about a grizzly's ways when left to himself. Besides what has been already said, we know that they appropriate game not killed by themselves, and will steal meat wherever it is found. Audubon saw one swimming in the Upper Missouri after the carcass of a drowned buffalo, Roosevelt had his elk eaten, and four of them visited Lord Dunraven's camp, carrying off all the food they could find. He says "they scarcely ate any of the flesh, but took the greatest pains to prevent any other creatures getting at it." This is not always the case, however. That they bury provisions is sure, but it is sometimes done very imperfectly, even when there is no physical difficulty in the way of completeness. On rocky soil the cache is simply covered with leaves, branches, and grass. Lord Dunraven, however, tells of a hunter who watched a grizzly burying its prey with the greatest care, concealing it completely, and finishing off his work in the most painstaking manner. Animals that have this habit need not watch their food as a tiger does his "kill," and when the interment was accomplished to this one's satisfaction, it went away. Before getting far, some "whiskey jacks" (a kind of magpie) that had been intently observing his doings began to unearth the deposit. Then he came back, drove them off, and repaired damages. This happened several times, until the bear flew into a violent passion, and while ramping around after the manner of these beasts he got shot. The author had a pony killed on one occasion, and the murderer buried its remains in the most slovenly manner possible.

These bears collect salmon during the spawning season on the banks of streams. They also scoop them out of the water with their claws, and dive after single fish. There are no full accounts of the manner in which prey is taken among these quadrupeds, but the creature's conformation makes it impossible that any of the deer kind could be captured except by stratagem. A grizzly can make a rapid rush. His lumbering, awkward gallop carries him forward so rapidly that on rough ground a man would have to be very fleet of foot to have any chance of escape. Colonel Markham states that the charge of an Indian hill bear is so swift that it cannot be avoided, and it appears from all accounts that so far as speed goes, at least for a short distance, the *Urside* have in general been underrated. In cover or upon open spaces, one of these bears always rises up when its attention is attracted, and it does the same if alarmed or angry, if wounded or intending to attack. It does this in order to see more clearly; for the sight, although it is not positively defective, cannot compare with that of many other species, and independently of the advantage gained by elevation, its short neck circumscribes vision while the body is in a horizontal position. The hearing is acute and the sense of smell highly developed. J. R. Bartlett, while acting upon the boundary commission between the United States and Mexico, says that at his encampment by the geysers of Pluton River his party found signs of these animals' proximity, but that they managed to avoid meeting the intruders, chiefly, as he supposed, by means of their scenting powers. Lieutenant J. W. Abert, while hidden with a companion at fifty yards from three grizzlies, was detected in this way, and the majority of observers have remarked upon the goodness of their noses. It is also said that they have an aversion to human effluvium, and that a warm trail will cause one to turn aside more certainly than the sight of a hunter. This needs confirmation, and may be taken with the same reservation which should attach to Godman's statement that the grizzly "is much more intimidated by the voice than the aspect of man." No doubt bears may have failed to push a charge home because their intended victim screamed with terror, but both in this case and in that just mentioned, while speaking of the influence of odor, so soon as such experiences are created into general truths, they can be met with facts by which they are stultified.

Nothing, so far as the author knows, has been advanced upon the subject of a male grizzly's paternal virtues or conjugal affections. As is the rule with fierce beasts, offspring depend upon the mother for care and protection. Two or three cubs are born together in spring, and they have been seen in her company from infancy up to an age when apparently able to shift for themselves. Very little is known, however, about the important subject of their training, the length of time during which they are under tutelage, or the degree to which tenderness and solicitude are developed in females of this species by maternity. A tigress robbed of her young has become a familiar simile for expressing desperation and inappeasable anger, but it has little foundation in truth, and many reports to the same effect in this animal's case, appear upon a wide survey of the evidence to be equally doubtful. Colonel R. I. Dodge ("Plains of the Great West") most likely comes as near the truth as it is possible for any one to do in the present state of knowledge, when he remarks that although a she-bear will often fight desperately in defence of her cubs, it is just as probable that they may be abandoned to their fate if the mother supposes herself to be in danger.

As might be imagined, grizzly bears can, for the most part, only be got the better of by being killed. They are occasionally trapped, however. The instrument is an ordinary toothed spring trap, to which a log is attached by a chain. When sprung it is impossible either to break or unloose it, and the furious animal goes off with the entire apparatus, but is much hampered by this encumbrance, and leaves a trail as easily followed as a turnpike.

Of necessity such a beast of prey as this has gathered around it a perfect fog of superstitions, traditions, false beliefs, and incredible stories. The author is familiar with the scenes in which most of these exploits and wonders are said to have been wrought, as well as with the men who relate and oftentimes believe them. As a class, they are not perhaps greatly superior in culture and mental discipline to those savages among whom their lives have been passed. Like them, their observations are generally accurate, and the inferences drawn from experience absurd. Travellers who associate with undeveloped men anywhere soon learn to make this distinction. Moreover, the trapper or hunter seen in general and most frequently met with in books, no more resembles some exceptional members of this class, than that blustering, melodramatic assassin, the would-be desperado, does the quiet, self-contained fighting-man of the frontier, and a wider difference than these classes present cannot be found among alien species in nature. If one is fortunate enough to find favor in the eyes of a true mountain man, he will do well to listen to what is said, and compare as many experiences with him as possible.

Among reports most rife upon the border is this, that if a fugitive pursued by a grizzly bear keeps a straight line around a hillside, the animal is certain to get either above or below him. The writer has heard men swear that they have tried this and seen it tried, but would be loath to trust in this device himself. Many persons are also convinced of the truth of a very prevalent account to the effect that a puma can kill one of these bears, and frequently does so. Nothing can be offered on the basis of personal experience or observation either in corroboration or rebuttal of this opinion. We have seen that there are good grounds for crediting the fact of Indian wild dogs assaulting tigers successfully, and the same is not impossible in this instance. Theodore Roosevelt ("Hunting Trips of a Ranchman") says "any one of the big bears we killed on the mountains would, I should think, have been able to make short work of a lion or a tiger." At the same time he remarks that either of the latter " would be fully as dangerous to a hunter or other human being, on account of the superior speed of its charge, the lightning-like rapidity of its movements, and its apparently sharper senses." The fact of an animal's antagonist being a man has evidently no relation to the question of relative prowess. Those advantages attributed to *Felidæ* must of course tell in conflict with any animal proportionately to the degree in which they exceeded like traits upon the part of an adversary. Cougars greatly excel the grizzly bear in those qualities mentioned, but how far they might counterbalance its great superiority in strength is another matter.

Nearly all that has been said of the subject of this sketch relates to his behavior towards human beings. Records of that character are not wanting, and it should be possible to give a correct idea of the grizzly as he appears in literature without overloading the text with quotations. Those traits to be considered in this connection are courage, ferocity, aggressiveness, and tenacity of life, all of which are represented very differently, according as the writers describe them from hearsay or personal observation, and as they refer to animals existing in dissimilar times and places, with or without reference to the fact that this is a creature which has undergone much modification under unlike conditions of existence. No one can delineate the features of this species in its entirety, but most persons attempt to do so, and their accounts are liable to the same objections which have been made to premature conclusions and want of discrimination in other instances.

The statements of those who know this animal do not disagree very conspicuously with respect to its character as a formidable foe. Dr. Elliott Coues, who, besides being a distinguished naturalist, had opportunities for acquiring a special knowledge of the grizzly bear, speaks of it in his "History of the Expedition of Lewis and Clark" in terms which afford a curious contrast to those of men who were less well informed. In mentioning the difficulties encountered by these explorers, he observes that "this bear was found to be so numerous and so fierce, especially in the upper Missouri region, as to more than once endanger the lives of the party, and form an impediment to the progress of the expedition." Lord Dunraven says that on "The Great Divide" these bears "did not appear to mind the proximity of our camp in the least, or to take any notice of us or our tracks. A grizzly is an independent kind of beast, and has a good deal of don't-care-a-damnativeness about him." Godman asserts that it is "justly considered to be the most dreadful and dangerous of American quadrupeds," while Audubon and Bachman, and, it may be added, the great majority of all who have had any personal acquaintance with the brute, refer to it in a similar way. Frederick Schwatka, for example, reports that "everywhere in his dismal dominions at the north he is religiously avoided by the native hunter. . . . Although he is not hunted, encounters with him are not unknown, as he is savage enough to become the hunter himself at times. . . . Indian fear of the great brown bear I found to be coextensive with all my travels in Alaska and the British Northwest Territory."

The other side in these opinions is represented by nobody more positively than Alfred G. Brehm ("Thierleben"). So far as one can judge from his work, he knew the animal of which he writes only by report, and if the text of his article is to be taken as an indication of the authorities consulted upon this subject, they were so few that it is not surprising he wandered far from reality. This author's views upon the character of *Ursus horribilis* may be thus given in English: "In its habits the gray bear is similar to ours; like these, it hibernates; but its walk is staggering and uncertain, and all its motions are heavier." Brehm states that in youth the grizzly climbs trees, that he is a good swimmer, "a thorough thief, and is strong enough to overpower every creature in his native country." When lassoed, he can drag up the horse. "Former writers have characterized him as a terrible and vicious animal that shows no fear of man, but, on the contrary, pursues him, whether mounted or on foot, armed or unarmed. . . . On all these grounds the hunter who has overcome Old Ephraim, as the bear is called, becomes the wonder and admiration of all mankind," including the Indians. "Among all their tribes the possession of a necklace of bears' claws and teeth gives its wearer a distinction which a prince or successful general scarcely enjoys among us." He must, however, have slain the animal from which these trophies were taken, himself. "Statements of this nature," remarks Brehm, "are some of them false and others greatly exaggerated. They were spread and believed at a time when the far West was but little visited, and when the public demanded an exciting story about a much dreaded animal that was fitted to play in the New World the same part that the famous beasts of prey did in the Old." This, with much more to the same effect ; and then, after a passing notice that Pechuel and Loesche found no grizzlies that would stand, he quotes General Marcy at length to show that they are rather harmless, cowardly, contemptible creatures, and dismisses the beast in disgrace.

Marcy relates ("Thirty Years of Army Life on the Border") that when he reached the haunts of grizzly bears, he expected to see destructive monsters in a perpetual rage, like Buffon's tigers. It was his belief that they would attack mounted men with rifles as soon as they came in sight, that these bears desired nothing more than to fight, in season and out of it, irrespective of time, place, or circumstances, and without reference to odds or any former experiences of the results. Not finding any such extraordinarily besotted idiots as this, the soldier, who seems to have been as fit to decide upon questions of comparative psychology as he was to give opinions in canon-law, became possessed with conceptions that are counterparts of those announced by Brehm. Those extracts made from the latter were taken from a very voluminous and undoubtedly valuable work on natural history, but its author has said nothing concerning the anomaly of a beast of prey twice as large as a lion and fully as well armed, being naturally timid and inoffensive, nor offered any suggestions with respect to those conditions which changed what must necessarily have been the brute's inherited character, before it began to avoid mankind; neither has he, apparently, taken more than the briefest glance at those accounts of the grizzly which give the results of personal observation. This animal is not customarily a hibernating one, it is not in the habit of climbing trees at any age, its reputation was far from being the outcome of a demand made by popular credulity. A grizzly bear could easily drag a horse up to him if he had hold of its riata. The Indian who killed one single-handed with a bow and arrows or trade-gun performed a feat second to none that can be imagined in the way of skill and daring, but thousands of rifle-carrying mountain men have done the like who took small credit to themselves, and got little from anybody else. This whole description is, considering its source, of the most surprising and unexpected character.

There are not many accounts of grizzly bears declining to fight; but it is evident that in this respect the animal, like every other beast that has been discussed, is more or less aggressive, according to the locality where it is found. Those bears Lewis and Clark encountered on the Upper Missouri in 1804, are like the grizzlies of the Yukon to-day, but their relations, that have been shot for nearly a century, know about rifles and conduct themselves accordingly. Theodore Roosevelt ("Still Hunting the Grizzly") expresses this change very well. "Now-a-days," he observes, "these great bears are much better aware than formerly of the death-dealing power of man, and, as a consequence, are far less fierce than was the case with their forefathers. . . . Constant contact with rifle-carrying hunters for a period extending over many generations of bear life, has taught the grizzly, by bitter experience, that man is his undoubted overlord, so far as fighting goes; and this knowledge has become a hereditary characteristic." With every advantage in arms, it is yet as dangerous to meet this brute fairly as to encounter a tiger on foot; and wherever that superiority has not been of long standing, grizzlies act like those that stalked Clark, charged Fremont, confronted Long, and killed Ross Cox's voyageur on the Columbia.

Colonel Dodge, referring to those that had become familiar with firearms, says that "a grizzly never attacks unless when wounded, or when he is cornered." This is, however, too general a statement. As one rides out of the Tejon Pass into the Tulare Valley, there is, a little to the right, an indentation or pocket in the foot-hills, in front of which stand some huge bowlders. From behind one of them a bear rushed out and destroyed the famous Andrew Sublette before he had an opportunity to defend himself. So far as that goes, the result might have been equally fatal if he had fired, for the writer used to carry his rifle, and it was far too light a weapon for such game as this. Goday, who was as renowned a paladin of the plains as he, related the circumstances of his death, and said that many similar cases had occurred in his experience. He added that one night, while sitting, as we were then, by the hearth of his little house at the mountain's base, there was a commotion outside at the corral, and going out in the darkness to see what was wrong, an immense bear rushed at him, and it was only by an instant that he got inside first. Many persons have been assailed by grizzly bears they never saw until too late, and the writer, except for the good fortune of being pitched over a precipice, would have been another. Some authors have a curious way of accounting for these incidents. They say that they occur because the animal was actually cornered, or if that statement cannot be made to fit the circumstances, its attack is attributed to an impression that it could not get away. There is no need to dwell upon this explanation. It is merely a blank assertion upon the part of those who know nothing about what the beast thinks or feels, and it is plainly one-sided in so far as it omits to take cognizance of the constitutional temper and tendencies of the creature whose acts are discussed.

No writer of any note except General Marcy has, so far as the author knows, denied that a grizzly bear soon comes to bay, and that he then devotes his energies to destruction with entire single-mindedness. Those who have met him, alike with those who have acquainted themselves with any completeness with the observations of others, know that this brute's patience under aggression is of the briefest, and his inherent ferocity easily aroused. When it is injured, the animal is exceptionally desperate, and fights from the first as a lion, tiger, and jaguar are apt to do only in their death rally. Colonel Dodge expresses the best opinions upon this point in saying that "when wounded, a grizzly bear attacks with the utmost ferocity, and regardless of the number or nature of his assailants. Then he is without doubt the most formidable and dangerous of wild beasts."

"In some way it has come about," says Lockwood,

"that . . . Bruin has secured for himself an almost superstitious respect." The way he did so has just been mentioned. Men had reason to fear him, and their veneration followed as a matter of course. It was because he proved "most formidable and dangerous" that Schwatka found among the Chilkat Indians the highest clan called brown bears, and for a like reason the native warrior wore his claws as a badge of honor.

Ferocity, prowess, and tenacity of life appear most conspicuously in accounts of actual conflict. Enough has been said with respect to the first-named trait, and no one ever called the others in question. Major Leveson ("Sport in Many Lands") is of the opinion that grizzly bears should only be met with the heaviest rifles — " bone-smashers," as Sir Samuel Baker calls them. Lighter weapons are too often ineffectual, and Dall ("Alaska and its Resources") reports that when the poorly armed natives of that province occasionally venture upon an assault of this kind, they assemble in large parties, watch the bear into the recesses of its den, block up the entrance with timber prepared for this purpose, and fire volleys into him as he tries to get at them. It will be denied by some, on anatomical grounds, that the Alaskan bears are grizzlies, but we are not concerned here with structural distinctions, and in character there is no difference. Colonel Dodge mentions the case of two soldiers at Fort Wingate who had an unfortunate encounter with one of these beasts, but does not give the details. Roosevelt, however, had the tale from the surgeon who attended them, and relates it ("Hunting Trips of a Ranchman") as follows: "The men were mail-carriers, and one day did not come in at the appointed time. Next day a relief party was sent out to look for them, and after some search found the bodies of both, as well as that of one of the horses. One of the men still showed signs of life; he came to his senses before dying, and told his story. They had seen a grizzly and pursued it on horseback, with their Spencer rifles. On coming close, one fired into its side, when it turned, with marvellous quickness for so large and unwieldy an animal, and struck down the horse, at the same time inflicting a ghastly wound upon the rider. The other man dismounted and came up to the rescue of his companion. The bear then left the latter and attacked him. Although hit by the bullet, it charged home and thrust the man down, and then lay on him and deliberately bit him to death, while his groans and cries were frightful to hear. Afterward it walked off into the bushes, without again offering to molest the already mortally wounded victim of his first assault."

It is commonly believed that feigning death will prevent a bear from inflicting further injuries. In many cases this is no doubt the case. Few unwounded animals tear a dead body, except in the act of devouring it. This stratagem must always be of doubtful efficacy, since beasts of prey would generally be acute enough to detect it. The ruse, however, may have been tried upon grizzlies with success; they are not brilliant beasts, so far as can be discovered; but this device sometimes fails. A hunter told the writer, over their camp-fire in the Sierra Nevada, of his brother's death, which he witnessed. They were shooting in those mountains, and he was on a steep escarpment of rock, his companion in the ravine beneath. A deer was roused and shot by the latter, when a large bear rushed upon him, struck the rifle out of his hands, and knocked him down, but without causing any serious injury. He said that he dared not fire for fear of infuriating the animal, and shouted to his brother to pretend to be dead. This was done; the beast walked round him, smelt at his body, and finally lay down close beside it. Suddenly he seized upon one of the arms and bit it savagely. The unfortunate man probably could not control respiration sufficiently, or there was some involuntary muscular movement. At all events, this is what happened, and the pain caused him to start up with a loud cry, upon which the bear rose erect, grasped him with his arms, and, in the language of the narrator, "bit the top of his head off clean."

Roosevelt relates that a neighbor of his, "out on a mining trip, was prospecting with two other men near the headwater of the Little Missouri, in the Black Hills country. They were walking down along the river, and came to a point of land thrust out into it, which was densely covered with brush and fallen timber. Two of the party walked round by the edge of the stream; but the third, a German, and a very powerful fellow, followed a well-beaten game trail leading through the bushy point. When they were some forty yards apart, these two men heard an agonized shout from the German, and at the same time the loud coughing growl or roar of a bear. They turned just in time to see their companion struck a terrible blow on the head by a grizzly, which must have been roused from its lair by his almost stepping on it; so close was it that he had no time to fire his rifle, but merely held it up over his head as a guard. Of course it was struck down, the claws of the great brute at the same time shattering his skull like an eggshell. Yet the man staggered on some ten feet before he fell; but when he did, he never spoke or moved again. The two others killed the bear after a short, brisk struggle, as he was in the midst of a most determined charge."

Everybody makes an oversight sometimes, and although this accomplished sportsman and careful writer is very free from the blemishes that usually disfigure observers of wild beasts, there is a slip of the pen here. How did he know this bear was not waiting for the man it killed? Nobody saw it until in the act of striking, and why the brute "*must* have been roused from its lair by his almost stepping upon it" does not appear. There is at least a probability that its acute senses warned it of the approach of a heavy man walking carelessly through brush, and of two others tramping round the cover within forty yards.

The bear's temper, disposition, and power of offence seem to be underrated with respect to the species at large. Whether because its appearance is less impressive than that of animals which have gathered about them most of the world's gossip, or for any other reason to which this inappreciation may be attributed, both in Europe, Asia, and America, the *Ursidæ* in general have undoubtedly less reputation than they seem to deserve, and less than the deeds they do and have done in all countries would apparently have brought with them as a matter of course. Poorly armed and primitive populations throughout the earth think differently, however, about them. In the folk-lore of Europe and Asia this creature is conspicuous. The great hunters write of it in a respectful strain. No man who ever stood before an enraged bear thought lightly of its prowess. A host of well-known names are appended to statements concerning destructive arctoids in the Scandinavian Mountains and the Pyrenees, in the Himalayas and Caucasus, the highlands of Central India, and the forests and plains north and south of "the stony girdle of the world."

There is every reason why this beast should be formidable wherever it has not encountered modern weapons; and that it is so its whole literature attests. Richardson's name ("Fauna Boreali Americana"), *Ursus ferox*, translates his own experiences and those of native tribes. Colonel Pollock ("Natural History Notes") asserts that "in Assam bears are far more destructive to human life than tigers," and more than one authoritative statement to the same effect has been made concerning those of India. It happens curiously that the ancient documents of China preserve the descriptive title which has been conferred upon the great bear of America. In Dr. Legge's edition of the Chinese Classics, the Bamboo Books have a note appended by some native scholiast to Part I., relating to the reign of Hwang-te, in which his general Ying-lung, while fighting against Ch'e-yew, is said to have been assisted by "tigers, panthers, bears, and *gristly* (*grizzly*) bears."

The grizzly is so difficult to kill that he has the reputation of being nearly invulnerable. It is quite true that the species possesses great tenacity of life, and that in extremity the animal is capable of doing extreme injury. "One of the most complete wrecks of humanity I ever saw," says Colonel Dodge, "was a man who had shot a grizzly bear through the head. Both were found dead together." Roosevelt killed one with a single shot. Following his trail among the Bighorn Mountains, he and his companion, while "in the middle of a thicket, crossed what was almost a breastwork of fallen logs, and Merrifield, who was leading, passed by the upright stem of a great pine. As soon as he was by it, he sank suddenly on one knee, turning half round, his face fairly aflame with excitement; and as I strode past him with my rifle at the ready, there was the great bear slowly rising from his bed among the young spruces. He had heard us . . . though we advanced with noiseless caution, . . . but apparently hardly knew exactly where or what we were, for he reared up on his haunches sideways to us. Then he saw us and dropped down again on all fours, the shaggy hair on his neck and shoulders seeming to bristle as he turned toward us. As he sank down on his fore feet, I raised the rifle; his head was slightly bent down, and when I saw the top of the white bead fairly between his small, glittering, evil eyes, I pulled trigger. Half rising up, the huge beast fell over on his side in the death throes, the ball having gone into his brain." Generally it is not so soon over. Captain Lewis mentions a case in which one did not succumb until eight balls went through its lungs, and several into other parts of the body. This officer also relates that one of his party was pursued for half a mile by a grizzly he had shot through the lungs, and which it finally took eight men to kill. Lewis said he would "rather encounter two Indians than one grizzly bear."

On the other hand, this powerful and ferocious creature may occasionally be destroyed or beaten off with seemingly inadequate means. Single Indians sometimes killed it; white hunters with "pea-rifles" often; and Roosevelt reports that he had a stallion that disabled one by a kick in the head. A similar account is given by Colonel Davidson ("Travels in Upper India") of an incurably vicious English thoroughbred at Lucknow, which fractured a tiger's skull when condemned to be devoured by this beast. Major Leveson, who had met most species of *Ursidæ*, regarded the grizzly as "by far the largest and most formidable of his race, . . . one of the most dangerous antagonists a hunter can meet with." But he knew that weapons before which the black rhinoceros and African elephant are powerless, prove too much for this animal also, and therefore refers "the numerous accidents that have occurred in hunting the grizzly to insufficiency of weight in the projectiles generally used." If the hunter be "armed with a large-bore breech-loading rifle, and keep his wits about him," he has the advantage, barring accident. But even then, "should the bear not be shot through the brain or heart, unless his assailant maintain his presence of mind, and put in his second barrel well and quickly, the chances are that the latter will come to grief, if his comrades fail to come to the rescue."

Leveson relates the following experience of his own: "We were encamped on the Wind River . . . when at daybreak one dreary morning a cry of alarm rang through camp, and I was awakened by our people hurrying to and fro in noisy confusion. . . . As I drew near to the clump of red cedars whence the sound of firearms issued . . . one of the half-breeds came running back and informed me that the row was occasioned by a grizzly, that had tried to carry off one of the baggage ponies, but had been driven off by the guard, who fired at him, and that in revenge he had carried off an Indian boy who had charge of the dogs. Guided by the shouting, which still continued, and accompanied by Pierre, who carried a second gun, I entered the copse and found a big grizzly evidently master of the situation; for although three or four of our Blackfoot scouts were halloaing around him, he did not appear to mind them, but confined his attentions to Crib, a bull-terrier, that pluckily kept him at bay by dancing about all round him, without risking a mauling by getting within striking reach of his claws. I was mounted on a thoroughly broken Indian mustang . . . and rode pretty close up before I saw that the boy was lying on the ground apparently so badly hurt as to be insensible, while the faithful old dog was doing what he could to protect him by harassing his huge antagonist.

"On my riding up to about twenty yards' distance, 'Old Ephraim' raised himself on his hind legs, and cocked his head knowingly on one side, as if he were going to make a rush. Whilst he was in this attitude, his brawny chest being fully exposed, I gave him the contents of both barrels almost simultaneously, which rolled him over on his back, where he made several convulsive movements with his paws. . . . Dismounting, I took my second gun from Pierre, and gave him the *coup de grâce* behind the ear, when, with a peculiarly melancholy, whining moan, he stretched out his great limbs and breathed his last."

The boy, though wounded, was feigning death and escaped, but it must be admitted that the ruse was tried under exceptionally favorable circumstances. "Many and many a spirit-stirring yarn," says Leveson, "have I heard related by hunters, around the watch-fire, of their encounters with the much-dreaded grizzly." Bear stories are greatly alike, he adds, and concludes his description by saying, in much the same way as Colonel Dodge ("The Black Hills"), "from my own experience, I should always give 'Old Ephraim' a wide berth if I were not armed with a thoroughly serviceable breech-loading rifle throwing a large ball."

The annals of hunting preserve the name of no greater or more adventurous sportsman than he who gives this opinion. It is one which every one who has encountered the grizzly bear will agree to, and it might also have been arrived at from studying the literature of this subject alone.

www.ingramcontent.com/pod-product-compliance
Lightning Source LLC
Chambersburg PA
CBHW030423300426
44112CB00009B/822